www.KnowledgePublications.com

DOE-HDBK-1015/1-93
JANUARY 1993

DOE FUNDAMENTALS HANDBOOK

CHEMISTRY

Volume 1 of 2

U.S. Department of Energy
Washington, D.C. 20585

FSC-6910

Distribution Statement A. Approved for public release; distribution is unlimited.

This document has been reproduced directly from the best available copy.

Available to DOE and DOE contractors from the Office of Scientific and Technical Information. P.O. Box 62, Oak Ridge, TN 37831; prices available from (615) 576-8401.

Available to the public from the National Technical Information Services, U.S. Department of Commerce, 5285 Port Royal., Springfield, VA 22161.

Order No. DE93011966

ABSTRACT

The *Chemistry* Handbook was developed to assist nuclear facility operating contractors in providing operators, maintenance personnel, and the technical staff with the necessary fundamentals training to ensure a basic understanding of chemistry. The handbook includes information on the atomic structure of matter; chemical bonding; chemical equations; chemical interactions involved with corrosion processes; water chemistry control, including the principles of water treatment; the hazards of chemicals and gases, and basic gaseous diffusion processes. This information will provide personnel with a foundation for understanding the chemical properties of materials and the way these properties can impose limitations on the operation of equipment and systems.

Key Words: Training Material, Atomic Structure of Matter, The Periodic Table of the Elements, Chemical Bonding, Corrosion, Water Chemistry Control, Water Treatment Principles, Chemical Hazards, Gaseous Diffusion Processes

FOREWORD

The *Department of Energy (DOE) Fundamentals Handbooks* consist of ten academic subjects, which include Mathematics; Classical Physics; Thermodynamics, Heat Transfer, and Fluid Flow; Instrumentation and Control; Electrical Science; Material Science; Mechanical Science; Chemistry; Engineering Symbology, Prints, and Drawings; and Nuclear Physics and Reactor Theory. The handbooks are provided as an aid to DOE nuclear facility contractors.

These handbooks were first published as Reactor Operator Fundamentals Manuals in 1985 for use by DOE category A reactors. The subject areas, subject matter content, and level of detail of the Reactor Operator Fundamentals Manuals were determined from several sources. DOE Category A reactor training managers determined which materials should be included, and served as a primary reference in the initial development phase. Training guidelines from the commercial nuclear power industry, results of job and task analyses, and independent input from contractors and operations-oriented personnel were all considered and included to some degree in developing the text material and learning objectives.

The *DOE Fundamentals Handbooks* represent the needs of various DOE nuclear facilities' fundamental training requirements. To increase their applicability to nonreactor nuclear facilities, the Reactor Operator Fundamentals Manual learning objectives were distributed to the Nuclear Facility Training Coordination Program Steering Committee for review and comment. To update their reactor-specific content, DOE Category A reactor training managers also reviewed and commented on the content. On the basis of feedback from these sources, information that applied to two or more DOE nuclear facilities was considered generic and was included. The final draft of each of the handbooks was then reviewed by these two groups. This approach has resulted in revised modular handbooks that contain sufficient detail such that each facility may adjust the content to fit their specific needs.

Each handbook contains an abstract, a foreword, an overview, learning objectives, and text material, and is divided into modules so that content and order may be modified by individual DOE contractors to suit their specific training needs. Each handbook is supported by a separate examination bank with an answer key.

The *DOE Fundamentals Handbooks* have been prepared for the Assistant Secretary for Nuclear Energy, Office of Nuclear Safety Policy and Standards, by the DOE Training Coordination Program. This program is managed by EG&G Idaho, Inc.

OVERVIEW

The *Department of Energy Fundamentals Handbook* entitled *Chemistry* was prepared as an information resource for personnel who are responsible for the operation of the Department's nuclear facilities. An understanding of chemistry will enable contractor personnel to understand the intent of the chemical concerns within their facility. A basic understanding of chemistry is necessary for DOE nuclear facility operators, maintenance personnel, and the technical staff to safely operate and maintain the facility and facility support systems. The information in the handbook is presented to provide a foundation for applying engineering concepts to the job. This knowledge will help personnel understand the impact that their actions may have on the safe and reliable operation of facility components and systems.

The *Chemistry* handbook consists of five modules that are contained in two volumes. The following is a brief description of the information presented in each module of the handbook.

Volume 1 of 2

Module 1 - Fundamentals of Chemistry

Introduces concepts on the atomic structure of matter. Discusses the periodic table and the significance of the information in a periodic table. Explains chemical bonding, the laws of chemistry, and chemical equations.

Appendix A - Basic Separation Theory

Introduces basic separation theory for the gaseous diffusion process. Discusses converter construction and basic operating principals.

Module 2 - Corrosion

Supplies basic information on the chemical interaction taking place during the corrosion process between the environment and the corroding metal.

OVERVIEW (Cont.)

Volume 2 of 2

Module 3 - Reactor Water Chemistry

Describes the chemical measures taken to retard the corrosion often found in water systems. The consequences of radioactivity on facility cooling water systems are also addressed.

Module 4 - Principles of Water Treatment

Details the principles of ion exchange in the context of water purity. Discusses typical water treatment methods and the basis for these methods.

Module 5 - Hazards of Chemicals and Gases

Explains why certain chemicals are considered hazardous to facility personnel. Includes general safety rules on handling and storage.

The information contained in this handbook is by no means all encompassing. An attempt to present the entire subject of chemistry would be impractical. However, the *Chemistry* Handbook does present enough information to provide the reader with a fundamental knowledge level sufficient to understand the advanced theoretical concepts presented in other subject areas, and to better understand basic system and equipment operation.

Department of Energy
Fundamentals Handbook

CHEMISTRY
Module 1
Fundamentals of Chemistry

TABLE OF CONTENTS

TABLE OF CONTENTS (Cont.)

TABLE OF CONTENTS (Cont.)

LIST OF FIGURES

LIST OF TABLES

REFERENCES

- Donald H. Andrews and Richard J. Kokes, Fundamental Chemistry, John Wiley & Sons, Inc., 1963
- Compressed Gas Association, Inc., Handbook of Compressed Gases, 2nd Edition, Reinhold Publishing Corporation, 1981.
- R. A. Day, Jr. and R. C. Johnson, General Chemistry, Prentice Hall, Inc., 1974.
- Dickerson, Gray, Darensbourg and Darensbourg, Chemical Principles, 4th Edition, The Benjamin Cummings Publishing Company, 1984.
- Academic Program for Nuclear Plant Personnel, Volume II, Chemistry, Columbia, MD, General Physics Corporation, Library of Congress Card #A 326517, 1972.
- General Physics Corporation, Fundamentals of Chemistry, General Physics Corporation, 1982.
- Glasstone and Sesonske, Nuclear Reactor Engineering, 3rd Edition, Van Nostrand Reinhold Company, 1981.
- McElroy, Accident Prevention Manual for Industrial Operations Engineering and Technology, Volume 2, 8th Edition, National Safety Council, 1980.
- Sienko and Plane, Chemical Principles and Properties, 2nd Edition, McGraw and Hill, 1974.
- Underwood, Chemistry for Colleges and Schools, 2nd Edition, Edward Arnold, Ltd., 1967.
- Norman V. Steere and Associates, CRC Handbook of Laboratory Safety, 2nd Edition, CRC Press, Inc., 1971.

TERMINAL OBJECTIVE

1.0 Without references, **DESCRIBE** the characteristics of an atom.

ENABLING OBJECTIVES

1.1 **DEFINE** the following terms:

a. States of matter
b. Atomic weight
c. Molecular weight
d. Mole
e. Gram atomic weight
f. Gram molecular weight

1.2 **LIST** the components of an atom, their relative sizes, and charges.

1.3 **STATE** the criterion used to classify an atom chemically.

1.4 **DEFINE** the following subdivisions of the periodic table:

a. Periods of the periodic table
b. Groups of the periodic table
c. Classes of the periodic table

1.5 Given a periodic table, **IDENTIFY** the following subdivisions:

a. Periods of the periodic table
b. Groups of the periodic table
c. Classes of the periodic table

1.6 **LIST** the characteristics that elements in the same group on the periodic table share.

1.7 **DEFINE** the term valence.

TERMINAL OBJECTIVE

2.0 Given an incomplete chemical equation, **BALANCE** the equation by the method presented.

ENABLING OBJECTIVES

2.1 **DEFINE** the following terms:

a. Ionic bonds
b. Van der Waals forces
c. Covalent bonds
d. Metallic bonds

2.2 **DESCRIBE** the physical arrangement and bonding of a polar molecule.

2.3 **DESCRIBE** the three basic laws of chemical reactions.

2.4 **STATE** how elements combine to form chemical compounds.

2.5 **EXPLAIN** the probability of any two elements combining to form a compound.

2.6 **DEFINE** the following terms:

a. Mixture
b. Solvent
c. Solubility
d. Solute
e. Solution
f. Equilibrium

2.7 **STATE** Le Chatelier's principle.

2.8 **DEFINE** the following terms:

a. ppm
b. Molarity
c. Density
d. Normality

2.9 **BALANCE** chemical equations that combine elements and/or compounds.

TERMINAL OBJECTIVE

3.0 Given sufficient information about a solution, **CALCULATE** the pH and pOH of the solution.

ENABLING OBJECTIVES

3.1 **DEFINE** the following terms:

a. Acid
b. Salt
c. pOH
d. Alkalies
e. Base
f. pH
g. Dissociation constant of water

3.2 **STATE** the formula for pH.

3.3 **STATE** the formula for pOH.

3.4 **CALCULATE** the pH of a specified solution.

Intentionally Left Blank

CHARACTERISTICS OF ATOMS

Chemistry is defined as the systematic investigation of the properties, structure, and behavior of matter and the changes matter undergoes. This general definition raises many questions. These questions are answered in the study of chemistry. Terms and basic concepts that help in understanding chemistry will be discussed in this chapter.

EO 1.1 DEFINE the following terms:

a.	**States of matter**	**d.**	**Mole**
b.	**Atomic weight**	**e.**	**Gram atomic weight**
c.	**Molecular weight**	**f.**	**Gram molecular weight**

EO 1.2 LIST the components of an atom, their relative sizes, and charges.

EO 1.3 STATE the criterion used to classify an atom chemically.

Characteristics of Matter

The term *states of matter* refers to the physical forms in which matter exists: solid, liquid, and gas. Solids are characterized as having both a definite shape and a definite volume. In a solid, the forces that keep the molecules or atoms together are strong. Therefore, a solid does not require outside support to maintain its shape.

Liquids have definite volumes but indefinite shapes and are slightly compressible. Liquids take the shape of their containers. The forces that keep a liquid's molecules or atoms together are weaker than in the solids.

Gases are readily compressible and capable of infinite expansion. They have indefinite shape and indefinite volume. Of the three states, gases have the weakest forces holding their molecules or atoms together.

The different states of matter have one thing in common; they can all be broken down into fundamental units called atoms.

The Atom Structure

All matter is composed of atoms, existing individually or in combination with each other. An atom is an extremely small electrically-neutral particle. It is the smallest unit involved in the chemical change of matter. Atoms can be treated as distinct particles because they behave as such chemically, but atoms themselves are composed of even smaller subparts. Understanding these atomic subparticles is important in understanding chemistry.

An atom is composed of a positively-charged nucleus orbited by one or more negatively-charged particles called electrons. A simplified schematic representation of this arrangement is illustrated in Figure 1. The nucleus is the core of an atom. It has a positive charge because it usually consists of two particles, the neutron and the proton (hydrogen is the exception with only a proton in the nucleus). The neutrons are electrically neutral, and the protons are electrically positive. A nucleus with one proton has a charge of +1 (or simply 1), and a nucleus with two protons has a +2 charge. Together the neutrons and protons give the nucleus its mass, but the proton alone gives the nucleus its positive charge.

Neutrons and protons are relatively massive and are essentially equal in mass.

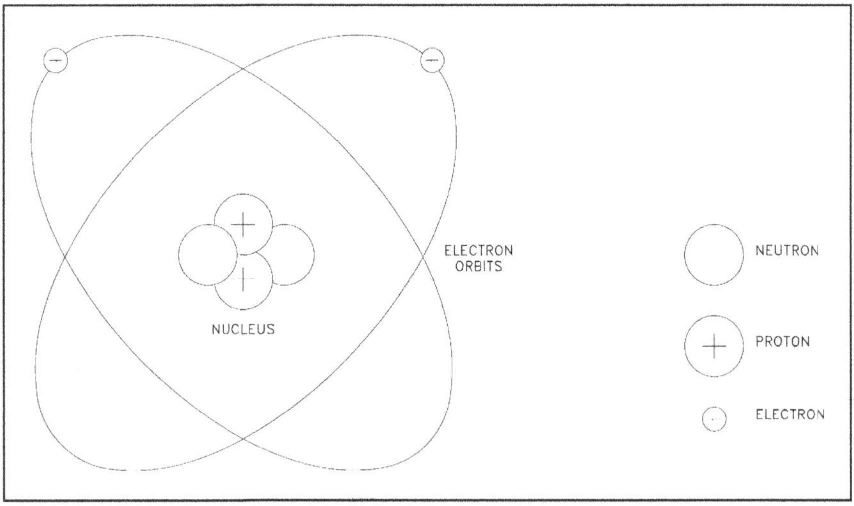

Figure 1 Schematic of a Simple Atom (Helium)

The particles that orbit the nucleus are electrons. They are very small, with a mass only 1/1835 the mass of a proton or neutron. Each electron is negatively charged, and the charge of one electron is equal in magnitude (but opposite in sign) to the charge of one proton. The number of electrons orbiting a nucleus is exactly equal to the number of protons contained in that nucleus. The equal and opposite charges cancel each other, and the atom as a whole is neutral. The electrons are bound in the atom by electrostatic attraction. The atom remains neutral unless some external force causes a change in the number of electrons.

The diameter of the atom is determined by the range of the electrons in their travels around the nucleus and is approximately 10^{-8} cm. The diameter of the nucleus is roughly 10,000 times smaller, approximately 10^{-13} to 10^{-12} cm. Because the nucleus is composed of neutrons and protons that are about 1835 times heavier than an electron, the nucleus contains practically all the mass of the atom, but constitutes a very small fraction of the volume. Although electrons are individually very small, the space in which they orbit the nucleus constitutes the largest part of the atomic volume.

Figure 1 illustrates these size relationships, but not to scale. If the nucleus were the size shown, the electrons would be several hundred feet away.

Some of the properties of the atom and its component parts are summarized in Table 1. The masses listed in Table 1 are measured in atomic mass units (amu), which is a relative scale in which the mass of a proton is about 1.0.

TABLE 1
Properties of the Atom and its Fundamental Particles

Particle Name	Relative Mass (amu)	Relative Charge (based on charge of proton)
Electron	0.00055 or 1/1835	-1
Proton	1.0	1
Neutron	1.0	0

Chemical Elements

An atom is classified chemically by the number of protons in its nucleus. Atoms that have the same number of protons in their nuclei have the same chemical behavior. Atoms that have the same number of protons are grouped together and constitute a chemical element.

Chemical Symbols

At one time chemists used various symbols, similar to shorthand, for the atoms of the different elements. These symbols were very cumbersome and were replaced by abbreviations of the names of the elements. Each element has been assigned a specific one or two letter symbol based on the first letter of its chemical name. Because there are several elements with the same first letter, it is often necessary to add the second letter to the symbol. In some cases the symbol comes from an abbreviation for the old latin name of the element. For example, Fe stands for iron (ferrum) and Cu for copper (cuprum). The first letter of the chemical symbol is always capitalized. If the symbol has two letters, the second letter is always lowercase.

Atomic Number

The number of protons in the nucleus plays such an important role in identifying the atom that it is given a special name, the *atomic number*. The symbol Z is often used for atomic number (or number of protons). Hydrogen has an atomic number of 1 and lawrencium has an atomic number of 103. The atomic number is also equal to the number of electrons.

Atomic Mass Number

The sum of the total number of protons, Z, and the total number of neutrons, N, is called the *atomic mass number*. The symbol is A. Not all atoms of the same element have the same atomic mass number, because, although the Z is the same, the N and thus the A are different. Atoms of the same element with different atomic mass numbers are called *isotopes*.

Atomic Weight

In Table 1, the masses of atomic particles are given in atomic mass units (amu). These units represent a relative scale in which the mass of the isotope carbon-12 is used as the standard and all others are related to it. Specifically, 1 amu is defined as 1/12 the mass of the carbon-12 atom. Since the mass of a proton or a neutron is approximately 1 amu, the mass of a particular atom will be approximately equal to its atomic mass number, Z.

The atomic weight of an element is generally more useful than isotopic masses. The *atomic weight* of an element is defined as the weighted average of the masses of all of its natural occurring isotopes. The atomic weight of the elements are listed in Table 2. The elements that have their atomic weights in parentheses are unstable. For these elements, the atomic weight of the longest living isotope is used rather than the average of the masses of all occurring isotopes.

TABLE 2
Table of Elements

Name and Symbol		Atomic Number	Atomic Weight (amu)	Name		Atomic Number	Atomic Weight (amu)
Actinium	Ac	89	(227)	Curium	Cm	96	(247)
Aluminum	Al	13	26.981	Dysprosium	Dy	66	162.50
Americium	Am	95	(243)	Einsteinium	Es	99	(252)
Antimony	Sb	51	121.75	Erbium	Er	68	167.26
Argon	Ar	18	39.948	Europium	Eu	63	151.96
Arsenic	As	33	74.921	Fermium	Fm	100	(257)
Astatine	At	85	(210)	Fluorine	F	9	18.998
Barium	Ba	56	137.34	Francium	Fr	87	(223)
Berkelium	Bk	97	(247)	Gadolinium	Gd	64	157.25
Beryllium	Be	4	9.012	Gallium	Ga	31	69.72
Bismuth	Bi	83	208.980	Germanium	Ge	32	72.59
Boron	B	5	10.811	Gold	Au	79	196.967
Bromine	Br	35	79.909	Hafnium	Hf	72	178.49
Cadmium	Cd	48	112.40	Helium	He	2	4.0026
Calcium	Ca	20	40.08	Holmium	Ho	67	164.930
Californium	Cf	98	(251)	Hydrogen	H	1	1.0079
Carbon	C	6	12.011	Indium	In	49	114.82
Cerium	Ce	58	140.12	Iodine	I	53	126.904
Cesium	Cs	55	132.905	Iridium	Ir	77	192.2
Chlorine	Cl	17	35.453	Iron	Fe	26	55.874
Chromium	Cr	24	51.996	Krypton	Kr	36	83.80
Cobalt	Co	27	58.933	Lanthanum	La	57	138.91
Copper	Cu	29	63.546	Lawrencium	Lw	103	(260)

TABLE 2 (Cont.)
Table of Elements

Name and Symbol		Atomic Number	Atomic Weight (amu)	Name		Atomic Number	Atomic Weight (amu)
Lead	Pb	82	207.19	Potassium	K	19	39.102
Lithium	Li	3	6.939	Praseodymium	Pr	59	140.90
Lutetium	Lu	71	174.97	Protactinium	Pa	91	231.03
Magnesium	Mg	12	24.312	Promethium	Pm	61	(145)
Manganese	Mn	25	54.938	Radium	Ra	88	226.02
Mendelevium	Md	101	(258)	Radon	Rn	86	(222)
Mercury	Hg	80	200.59	Rhenium	Re	75	186.2
Molybdenum	Mo	42	95.94	Rhodium	Rh	45	102.90
Neodymium	Nd	60	144.24	Rubidium	Rb	37	85.47
Neon	Ne	10	20.183	Ruthenium	Ru	44	101.07
Neptunium	Np	93	237.05	Samarium	Sm	62	150.35
Nickel	Ni	28	58.71	Scandium	Sc	21	44.956
Niobium	Nb	41	92.906	Selenium	Se	34	78.96
Nitrogen	N	7	14.006	Silicon	Si	34	78.96
Nobelium	No	102	(259)	Silver	Ag	47	107.87
Osmium	Os	76	190.2	Sodium	Na	11	22.989
Oxygen	O	8	15.999	Strontium	Sr	38	87.62
Palladium	Pd	46	106.41	Sulfur	S	16	32.064
Phosphorus	P	15	30.973	Tantalum	Ta	73	180.94
Platinum	Pt	78	195.09	Technetium	Tc	43	(98)
Plutonium	Pu	94	(244)	Tellurium	Te	52	127.60
Polonium	Po	84	(209)	Terbium	Tb	65	158.92

TABLE 2 (Cont.)
Table of Elements

Name and Symbol		Atomic Number	Atomic Weight (amu)	Name		Atomic Number	Atomic Weight (amu)
Thallium	Tl	81	204.37	Vanadium	V	23	50.942
Thorium	Th	90	232.03	Xenon	Xe	54	131.30
Thulium	Tm	69	168.93	Ytterbium	Yb	70	173.04
Tin	Sn	50	118.69	Yttrium	Y	39	88.905
Titanium	Ti	22	47.90	Zinc	Zn	30	65.37
Tungsten	W	74	183.85	Zirconium	Zr	40	91.22
Uranium	U	92	238.03				

Molecules

Molecules are groups or clusters of atoms held together by means of chemical bonding. There are two types of molecule; molecules of an element and molecules of a compound.

Molecules of an Element

In certain cases, two single atoms of an element can be attracted to one another by a bond to form a molecule. Examples of this are hydrogen, oxygen, and bromine. The molecular formulas for these are H_2, O_2, and Br_2. Most gaseous elements exist as molecules of two atoms.

Molecules of a Compound

Two atoms of different elements held together by a bond form a compound. The molecule is the primary particle of a chemical compound. Some examples of this type of molecule include hydrogen chloride (HCl), water (H_2O), methane (CH_4), and ammonia (NH_3).

Molecular Weight

The weight of a molecule, the *molecular weight,* is the total mass of the individual atoms. Therefore, it is fairly simple to calculate the mass of any molecule if its formula is known (that is, the elements and the number of each that make up the molecule). Note that the terms mass and weight are used interchangeably in chemistry.

Example 1:

The compound water has a formula of H_2O. This means there is one atom of oxygen and two atoms of hydrogen. Calculate the molecular weight.

Solution:

The molecular weight is calculated as follows:

1 atom	×	16.000	(the atomic weight of oxygen)	=	16.000 amu
2 atoms	×	1.008	(the atomic weight of hydrogen)	=	2.016 amu
			molecular weight of water	=	18.016 amu

Example 2:

Calculate the molecular weight of H_2SO_4.

Solution:

hydrogen	2 atoms	×	1.008 amu	=	2.016 amu
sulfur	1 atom	×	32.064 amu	=	32.064 amu
oxygen	4 atoms	×	15.999 amu	=	63.996 amu
			molecular weight	=	98.076 amu

Example 3:

Calculate the molecular weight of HCl.

Solution:

hydrogen	1 atom	×	1.008 amu	=	1.008 amu
chlorine	1 atom	×	35.453 amu	=	35.453 amu
			molecular weight	=	36.461 amu

Avogadro's Number

Consider one atom of oxygen and one atom of sulfur, and compare their atomic weights.

Oxygen's atomic weight = 15.999 amu

Sulfur's atomic weight = 32.06 amu

The sulfur atom weighs approximately twice as much as the oxygen atom.
$(32.06 \div 15.99 \cong 2)$

Because the sulfur atom weighs twice as much as an oxygen atom, a one gram sample of oxygen contains twice as many atoms as a one gram sample of sulfur. Thus, a two gram sample of sulfur contains the same number of atoms as a one gram sample of oxygen.

From this previous example, one might suggest that a relationship exists between the weight of a sample and the number of atoms in the sample. In fact, scientists have determined that there is a definite relationship between the number of atoms in a sample and the sample's weight. Experimentation has shown that, for any element, a sample containing the atomic weight in grams contains 6.022×10^{23} atoms. Thus 15.999 grams of oxygen contains 6.022×10^{23} atoms, and 32.06 grams of sulfur contains 6.022×10^{23} atoms. This number (6.022×10^{23}) is known as *Avogadro's number*. The importance of Avogadro's number to chemistry should be clear. It represents the number of atoms in X grams of <u>any</u> element, where X is the atomic weight of the element. It permits chemists to predict and use exact amounts of elements needed to cause desired chemical reactions to occur.

The Mole

A single atom or a few atoms are rarely encountered. Instead, larger, macroscopic quantities are used to quantify or measure collections of atoms or molecules, such as a glass of water, a gallon of alcohol, or two aspirin. Chemists have introduced a large unit of matter, the mole, to deal with macroscopic samples of matter.

One mole represents a definite number of objects, substances, or particles. (For example, a mole of atoms, a mole of ions, a mole of molecules, and even, theoretically, a mole of elephants.) A *mole* is defined as the quantity of a pure substance that contains 6.022×10^{23} units (atoms, ions, molecules, or elephants) of that substance. In other words, a mole is Avogadro's number of anything.

For any element, the mass of a mole of that element's atoms is the atomic mass expressed in units of grams. For example, to calculate the mass of a mole of copper atoms, simply express the atomic mass of copper in units of grams. Because the atomic mass of copper is 63.546 amu, a mole of copper has a mass of 63.546 grams. The value for the atomic mass of gold is 196.967 amu. Therefore, a mole of gold has a mass of 196.967 grams. The mass of a mole of atoms is called the *gram atomic weight* (GAW). The mole concept allows the conversion of grams of a substance to moles and vice versa.

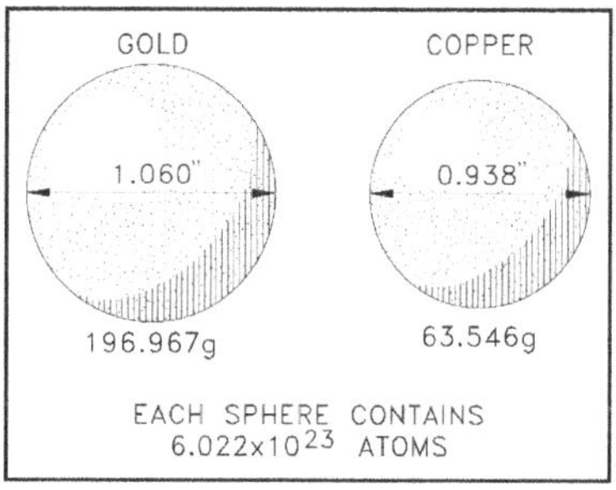

Figure 2 A Mole of Gold Compared to a Mole of Copper

Figure 2 contains a ball of gold and a ball of copper. The two balls are of different masses and different sizes, but each contains an identical number of atoms.

Example 1:

A silver bar has a mass of 1870 grams. How many moles of silver are in the bar?

Solution:

Since the atomic mass of silver (Ag) is 107.87 amu, one mole of silver has a mass 6 107.87 grams. Therefore, there is one mole of Ag per 107.87 grams of Ag o $\frac{1 \text{ mole Ag}}{107.87 \text{ grams Ag}}$. There are 1870 grams of silver.

$$\frac{1870 \text{ grams Ag}}{1} \times \frac{1 \text{ mole Ag}}{107.87 \text{ grams Ag}} = 17.3 \text{ mole Ag}$$

Example 2:

Mercury (Hg) is the only metal that exists as a liquid at room temperature. It isused in thermometers. A thermometer contains 0.004 moles of mercury. How many grams 6 mercury are in the thermometer?

Solution:

Since the atomic mass of Hg is 201 amu, one mole of Hg has a mass of 201 grams of Hg or $\frac{201 \text{ grams Hg}}{1 \text{ mole Hg}}$. There are 0.004 moles of Hg.

$$\frac{0.004 \text{ moles Hg}}{1} \times \frac{201 \text{ grams Hg}}{1 \text{ mole Hg}} = 0.8 \text{ grams Hg}$$

Mole of Molecules

The mass of a mole of molecules ofa substance is the molecular mass expressed in grams. For example, an oxygen molecule (O_2) has a molecular mass equivalent to 32.0 grams because each oxygen atom has a molecular mass of 16.0 grams. (Recall that to obtain the molecular mass, the atomic masses of all atoms that appear in the formula are added.) If the atomic masses of th carbon and four hydrogen atoms in methane, CH_4, are added, the total is 16 amu. Therefore, one mole of CH_4 has a mass of 16 grams. The mass of a mole of molecules is called the molar mass or *gram molecular weight* (GMW).

Summary

The important information found in this chapter is summarized below.

Characteristics of Atoms Summary

- The following terms are defined:

  States of matter is a term which refers to the physical forms in which matter exists; solid, liquid and gas. Solids are characterized as having both a definite shape and a definite volume. Liquids have definite volumes but indefinite shapes and are slightly compressible. Gases are readily compressible and capable of infinite expansion.

  Atomic weight is defined as the weighted average of the masses of all its natural occurring isotopes.

  Molecular weight will be the total weight of the individual atoms of a molecule.

  A mole is Avogadro's number of any substance.

  Gram atomic weight is the mass of a mole of atoms.

  Gram molecular weight is the mass of a mole of molecules, (GMW).

- The components of an atom are protons, neutrons, and electrons. A proton has a mass of 1.0 amu and a positive charge (+1). The neutron also has a mass of 1.0 amu but is neutral in charge. The electron has a mass of .00055 or 1/1835 amu and a negative charge (-1).

- An atom is classified chemically by the number of protons in its nucleus.

THE PERIODIC TABLE

All known elements fall into a pattern when placed in a periodic table, and the position in this pattern is decided by the element's atomic number. This chapter will discuss the significance of this fact.

EO 1.4 DEFINE the following subdivisions of the periodic table:

a. Periods of the periodic table
b. Groups of the periodic table
c. Classes of the periodic table

EO 1.5 Given a periodic table, IDENTIFY the following subdivisions:

a. Periods of the periodic table
b. Groups of the periodic table
c. Classes of the periodic table

EO 1.6 LIST the characteristics that elements in the same group on the periodic table share.

EO 1.7 DEFINE the term valence.

Periodic Table

Over many years of chemical investigation, scientists have discovered a remarkable feature of the elements. If the elements are arranged in the order of their atomic numbers, the chemical properties of the elements are repeated somewhat regularly. To a lesser extent, the physical properties are also repeated periodically. This periodic repetition can be seen in Table 3. Compare the properties of lithium (Li), sodium (Na), and potassium (K), and also those of beryllium (Be), magnesium (Mg), and calcium (Ca). In the list of elements shown in Table 3 the properties are repeated every eighth element.

TABLE 3
Description of the Properties of the First Twenty Elements

Element	Symbol	Atomic Number	Atomic Weight	Description of Properties
Hydrogen	H	1	1.008	Colorless gas, reacts readily with oxygen to form H_2O; forms HCl with chlorine.
Helium	He	2	4.003	Colorless gas, very non-reactive chemically.

TABLE 3 (Cont.)
Description of the Properties of the First Twenty Elements

Lithium	Li	3	6.939	Silvery white, soft metal, very reactive chemically, forms Li_2O and LiCl readily.
Beryllium	Be	4	9.012	Grey metal, much harder than lithium, fairly reactive chemically, forms BeO and $BeCl_2$ easily.
Boron	B	5	10.811	Yellow or brown non-metal, very hard element, not very reactive, but will form B_2O_3, and BCl_3
Carbon	C	6	12.011	Black non-metal, brittle, non-reactive at room temperature. Forms CO_2 and CCl_4.
Nitrogen	N	7	14.007	Colorless gas, not very reactive, will form N_2O_5 and NH_3.
Oxygen	O	8	15.999	Colorless gas, moderately reactive,will combine with most elements, forms CO_2, MgO, etc.
Fluorine	F	9	18.998	Green-yellow gas, extremely reactive, irritating to smell, forms NaF, MgF_2.
Neon	Ne	10	20.183	Colorless gas, very non-reactive chemically.
Sodium	Na	11	22.990	Silvery soft metal, very reactive chemically, forms Na_2O and NaCl.
Magnesium	Mg	12	24.312	Silvery white metal, much harder than sodium. Fairly reactive, forms MgO and $MgCl_2$.
Aluminum	Al	13	26.982	Silvery white metal, like magnesium but not as reactive. Forms Al_2O_3 and $AlCl_3$.
Silicon	Si	14	28.086	Gray, non-metallic, non-reactive at room temperature, forms SiO_2 and $SiCl_4$.
Phosphorus	P	15	30.974	Black, red, violet, or yellow solid, low melting point, quite reactive, forms P_2O_5 and PCl_3.
Sulfur	S	16	32.064	Yellow solid with low melting point. Moderately reactive, combines with most elements, forms SO_2, MgS, etc.
Chlorine	Cl	17	35.453	Greenish-yellow gas, extremely reactive, irritating to smell, forms NaCl, $MgCl_2$.
Argon	Ar	18	39.948	Colorless gas, very non-reactive chemically.
Potassium	K	19	39.102	Silver soft metal, very reactive chemically, forms K_2O and KCl.
Calcium	Ca	20	40.080	Silver-white metal, much harder than potassium, fairly reactive, forms CaO and $CaCl_2$.

A table in which elements with similar chemical properties are grouped together is called a periodic table. One of the most common versions is shown in Figure 3. In this table, elements are arranged in order of increasing atomic number in succeeding rows. Each horizontal row is called a *period*. Note that some periods are longer than others. Elements with similar chemical properties appear in vertical columns called *groups*. Each group is designated by a Roman numeral and a capital letter, except the one on the extreme right-hand side, Group 0 (the inert gases). At the bottom of the periodic table are two long rows of elements identified as the lanthanide series and the actinide series. They are separated from the table primarily to keep it from becoming too wide. Also, the elements within each of these two series show similar chemical properties.

The number directly below each element is its atomic number, and the number above each element is its atomic weight. In several cases the atomic weights are in parentheses. This indicates that these elements have no stable isotopes; that is, they are radioactive. The value enclosed in parentheses and used for the atomic weight is the atomic mass number of the most stable known isotope, as indicated by the longest half-life.

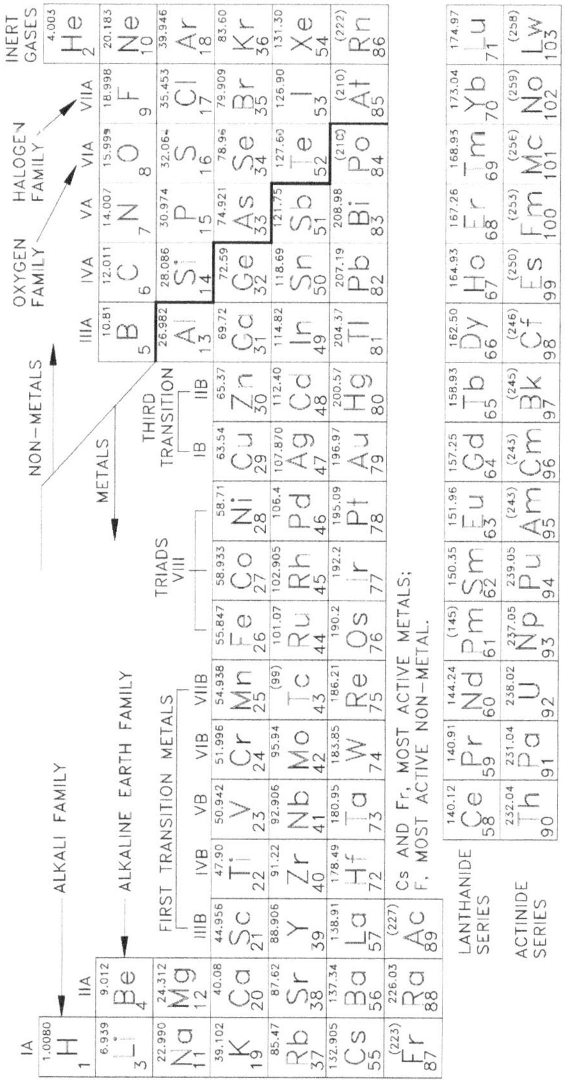

Figure 3 Periodic Table of the Elements

Classes of the Periodic Table

There are three broad *classes* of elements. These are the metals, the non-metals, and the semi-metals. These three classes are grouped together on the periodic table as shown on Figure 4.

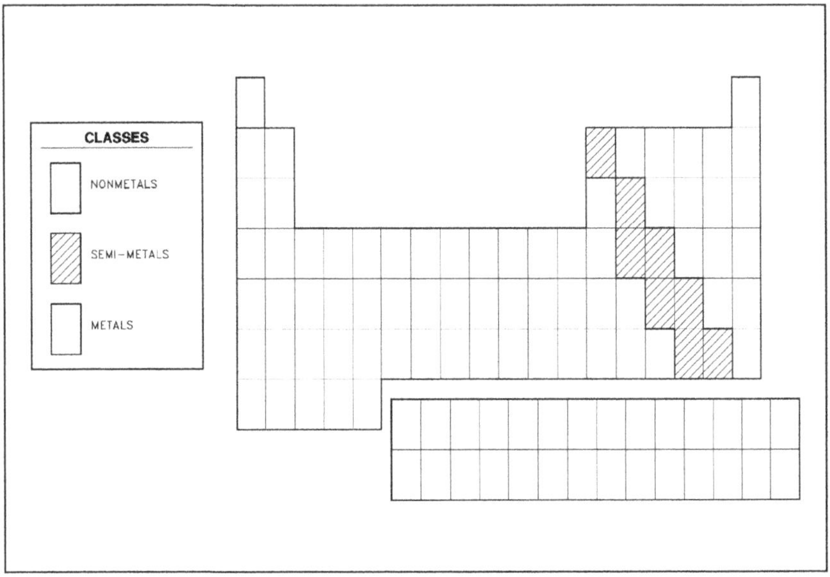

Figure 4 Regional Schematic of Periodic Table

Metals

The metals constitute the largest class of elements and are located on the left and toward the center of the periodic table as shown in Figure 4. In Figure 3, the heavy line running step-wise from boron (B) to astatine (At) generally separates the metals from the rest of the elements (elements in the actinide and lanthanide series are metals). Metals tend to lose electrons to form positive ions rather than to gain electrons and become negative ions.

Most people are familiar with metals' physical properties. They are usually hard and strong, capable of being shaped mechanically (malleable and ductile), and good conductors of heat and electricity, and they have lustrous surfaces when clean. More important for chemical classification are the chemical properties of metals because the physical properties are not common to all metals. For example, mercury (Hg) is a metal, although it is a liquid at room temperature, and sodium is a metal although it is not at all hard or strong. Metals can be involved in a wide range of chemical reactions. Their reactions with water range from violent with sodium and potassium to imperceptible with gold and platinum. Metals are divided into the following two categories.

1. The light metals, which are soft, have a low density, are very reactive chemically, and are unsatisfactory as structural materials.

2. The transition metals, which are hard, have a high density, do not react readily, and are useful structural materials.

The metals in Category 1 are located at the far left of the table (Groups IA and IIA). The metals in Category 2 are located in the middle of the table (the B groups).

Nonmetals

The nonmetals occupy the part of the periodic table to the right of the heavy, step-like line. (refer to Figure 3 and Figure 4)

In general, the physical properties of the nonmetals are the opposite of those attributed to metals. Nonmetals are often gases at room temperature. The nonmetals that are solids are not lustrous, are not malleable or ductile, and are poor conductors of heat and electricity. Some nonmetals are very reactive, but the nature of the reactions is different from that of metals. Nonmetals tend to gain electrons to form negative ions rather than to lose electrons to form positive ions.

The six elements in Group 0 represent a special subclass of nonmetals. They are all very unreactive gases, so they are called the inert gases. For many years it was believed that the inert gases would not and could not participate in chemical reactions. In 1962, the first true compounds of an inert gas, XeF_4 and $XePtF_6$, were positively identified.

Since that time, several other compounds have been prepared. The preparation of these compounds requires special conditions; under ordinary conditions, the inert gases may be considered nonreactive.

Semi-Metals

The obvious trend in the periodic table is that from left to right, across any period, the elements change from distinctly metallic (Group IA) to distinctly nonmetallic (Group VIIA). This change in character is not sharply defined, but is gradual. Generally, elements well to the left of the heavy diagonal line are metals, and those well to the right are nonmetals. Some of the elements near the line, however, exhibit properties of metals under some conditions and properties of nonmetals under other conditions. These elements are called the semi-metals and include boron (B), silicon (Si), germanium (Ge), arsenic (As), and tellurium (Te). They are usually classified as semi-conductors of electricity and are widely used in electrical components.

Group Characteristics

Each set of elements appearing in the vertical column of a periodic table is called a Group and represents a family of elements that have similar physical and chemical properties. Group IA is the Alkali Family; Group IIA is the Alkaline Earth Family; Group VIA is the Oxygen Family; Group VIIA is the Halogen Family. On the left side of the table are Group IA elements (except hydrogen), which are soft metals that undergo similar chemical reactions. The elements in Group IIA form similar compounds and are much harder than their neighbors in Group IA.

As shown in the previous section, there are some exceptions to the generalizations concerning chemical properties and the periodic table. The most accurate observation is that all elements within a particular group have similar physical and chemical properties.

This observation is most accurate at the extreme sides of the table. All elements in Group 0 are unreactive gases, and all elements in Group VIIA have similar chemical properties, although there is a gradual change in physical properties. For example, fluorine (F) is a gas while iodine (I) is a solid at room temperature.

Groups with a B designation (IB through VIIB) and Group VIII are called transition groups. In this region of the table, exceptions begin to appear. Within any group in this region, all the elements are metals, but their chemical properties may differ. In some cases, an element may be more similar to neighbors within its period than it is to elements in its group. For example, iron (Fe) is more similar to cobalt (Co) and nickel (Ni) than it is to ruthenium (Ru) and osmium (Os). Most of these elements have several charges, and their ions in solution are colored (ions of all other elements are colorless).

The line separating metals from nonmetals cuts across several groups. In this region of the table, the rule of group similarities loses much of its usefulness. In Group IVA, for example, carbon (C) is a nonmetal; silicon (Si) and germanium (Ge) are semi-metals; and tin (Sn) and lead (Pb) are metals.

Chemical activity can also be determined from position in the periodic table. The most active metals are the members of the Alkali Family, e.g., cesium (Cs) and francium (Fr). The most active nonmetals are the members of the Halogen Family, e.g., fluorine (F) and chlorine (Cl). The Noble Gases in Group 0 are inert. The activity of metals decreases when proceeding to the right in the periodic table; the activity of nonmetals decreases when proceeding to the left.

Atomic Structure of Electrons

Based on experimental data, it is known that chemical reactions involve only the electrons in atoms. In fact, only some of the electrons are involved. Because chemical properties are periodic, there must also be a periodic characteristic about electrons. This characteristic is the manner in which electrons are arranged in the atom. Electrons are in constant motion around the nucleus. They have both kinetic and potential energy, and their total energy is the sum of the two. The total energy is quantized; that is, there are definite, discrete values of total energy that atomic electrons can possess. These energy states can be visualized as spherical shells around the nucleus separated by forbidden areas where electrons cannot exist in a stable state. This sort of arrangement is illustrated in Figure 5.

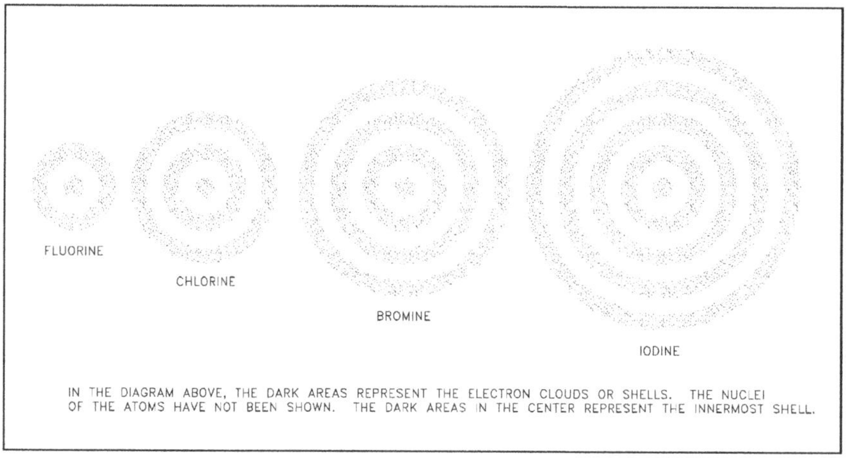

Figure 5 Electron Shells of Atoms

It is customary to speak of electron shells around the nucleus, and the shells are referred to by number. The first, or No. 1, shell is the one nearest the nucleus; the second, or No. 2, shell is next; then the third, or No. 3, shell; and so on in numerical order. In general, electrons closer to the nucleus have a lower energy state. Atomic electrons always seek the lowest energy state available.

The electron shells represent major energy states of electrons. Each shell contains one or more subshells called orbitals, each with a slightly different energy. In order of increasing energy, the orbitals are designated by the small letters s, p, d, f, g, h.

No two shells consist of the same number of orbitals. The first shell contains only one orbital, an s orbital. The second shell contains s and p orbitals. In general, each higher shell contains a new type of orbital:

> the first shell contains an s orbital,
> the second shell contains s and p orbitals,
> the third shell contains s, p, and d orbitals,
> the fourth shell contains s, p, d, and f orbitals,

and so on. Each orbital can hold a definite maximum number of electrons. There is also a limit to the number of electrons in each shell and the limit increases as one goes to higher shells. The numbers of electrons that can occupy the different orbitals and shells are shown in Table 4.

TABLE 4
Electrons, Orbital, and Shell Relationships in Atomic Structure

Shell Number	Type of Orbitals	Maximum Number of Electrons in Each Orbital	Maximum Total Electrons in shell
1	s	2	2
2	s	2	8
	p	6	
3	s	2	18
	p	6	
	d	10	
4	s	2	32
	p	6	
	d	10	
	f	14	
5	s	2	50
	p	6	
	d	10	
	f	14	
	g	18	

A more specific statement can now be made about which electrons are involved in chemical reactions. Chemical reactions involve primarily the electrons in the outermost shell of an atom. The term outermost shell refers to the shell farthest from the nucleus that has some or all of its allotted number of electrons. Some atoms have more than one partially-filled shell. All of the partially-filled shells have some effect on chemical behavior, but the outermost one has the greatest effect. The outermost shell is called the valence shell, and the electrons in that shell are called valence electrons. The term *valence* (of an atom) is defined as the number of electrons an element gains or loses, or the number of pairs of electrons it shares when it interacts with other elements.

The periodic chart is arranged so that the valence of an atom can be easily determined. For the elements in the A groups of the periodic chart, the number of valence electrons is the same as the group number; that is, carbon (C) is in Group IVA and has four valence electrons. The noble gases (Group 0) have eight in their valence shell with the exception of helium, which has two.

The arrangement in which the outermost shell is either completely filled (as with He and Ne) or contains eight electrons (as with Ne, Ar, Kr, Xe, Rn) is called the inert gas configuration. The inert gas configuration is exceptionally stable energetically because these inert gases are the least reactive of all the elements.

The first element in the periodic table, hydrogen, does not have properties that satisfactorily place it in any group. Hydrogen has two unique features: (a) the highest energy shell of a hydrogen atom can hold only two electrons, in contrast to all others (except helium) that can hold eight or more; and (b) when hydrogen loses its electron, the ion formed, H^+, is a bare nucleus. The hydrogen ion is very small in comparison with a positive ion of any other element, which must still have some electrons surrounding the nucleus. Hydrogen can either gain or lose an electron. It has some properties similar to Group IA elements, and some similar to Group VIIA elements.

The number of electrons in the outer, or valence, shell determines the relative activity of the element. The elements are arranged in the periodic table so that elements of the same group have the same number of electrons in the outer shell (except for the Transition Groups). The arrangement of electrons in the outer shell explains why some elements are chemically very active, some are not very active, and others are inert. In general, the fewer electrons an element must lose, gain, or share to reach a stable shell structure, the more chemically active the element is. The likelihood of elements forming compounds is strongly influenced by this valence shell and on the stability of the resulting molecule. The more stable the molecules are, the more likely these molecules are to form.

Summary

The important information from this chapter is summarized below.

Periodic Table Summary

- The subdivisions of the periodic table are periods, groups, and classes. The horizontal rows are called periods. The vertical columns are called groups. The entire table consists of three classes: metals, non-metals, and semi-metals.

- The subdivisions of the periodic chart have been explained such that the student should be able to identify them if given a periodic table.

- Elements of the same group share certain physical and chemical characteristics. Examples of the characteristics of several groups are listed below.

  Group 0 contains elements that are unreactive gases.
  Group IA contains elements that are chemically active soft metals.
  Group VIIA contains elements that are chemically active nonmetals.
  Groups IB through VIIB and VIII are called transition groups and elements found in them display properties of metals.

- The *valence* of an atom is defined as the number of electrons an element gains or loses, or the number of pairs of electrons it shares when it interacts with other elements.

CHEMICAL BONDING

The development of matter, no matter what the form, is the result of the practical application of the assumptions, hypotheses, theories, and laws that chemists have formulated from their research into the nature of matter, energy, and change. This chapter will address some of these theories and laws. Chemical bonds and how atoms bond to form molecules will be discussed. In addition, an introduction to organic chemistry is provided.

EO 2.1 DEFINE the following terms:

a.	**Ionic bonds**	**c.**	**Covalent bonds**
b.	**Van der Waals forces**	**d.**	**Metallic bonds**

EO 2.2 DESCRIBE the physical arrangement and bonding of a polar molecule.

EO 2.3 DESCRIBE the three basic laws of chemical reactions.

EO 2.4 STATE how elements combine to form chemical compounds.

EO 2.5 EXPLAIN the probability of any two elements combining to form a compound.

EO 2.6 DEFINE the following terms:

a.	**Mixture**	**d.**	**Solute**
b.	**Solvent**	**e.**	**Solution**
c.	**Solubility**	**f.**	**Equilibrium**

Chemical Bonding

As stated in the previous chapter, the number of electrons in the outer, or valence, shell determines the relative activity of the element. The arrangement of electrons in the outer shell explains why some elements are chemically very active, some are not very active, and others are inert. In general, the fewer electrons an element must lose, gain, or share to reach a stable shell structure, the more chemically active the element is. The likelihood of elements forming compounds is strongly influenced by the completion of the valence shell and by the stability of the resulting molecule. The more stable the resulting molecules are, the more likely these molecules are to form. For example, an atom that "needs" two electrons to completely fill the valence shell would rather react with another atom which must give up two electrons to satisfy its valence.

In the case of $H^+ + Br^-$, this is likely to take place because the exchange would satisfy the needs of both atoms. Although there is far more to consider than just the number of valence electrons, this is a good rule of thumb.

If the atom needed two electrons and only picked up one, it would still actively seek out an additional electron. The reaction of $H^+ + Te^{-2}$ is far less likely to take place because the resulting molecule would still have an incomplete valence shell. Of course, the combining of two atoms, when both want to release or gain electrons, may take place (for example; H_2 or O_2) but is less probable when other atoms are available.

Atoms are joined or bonded together through this interaction of their electrons. There are several types of chemical bonds that hold atoms together; three will be discussed, ionic, covalent, and metallic.

Ionic Bonds

An *ionic bond* is formed when one or more electrons is wholly transferred from one element to another, and the elements are held together by the force of attraction due to the opposing charges. An example of ionic bonding is shown in Figure 6(A) for sodium chloride (table salt).

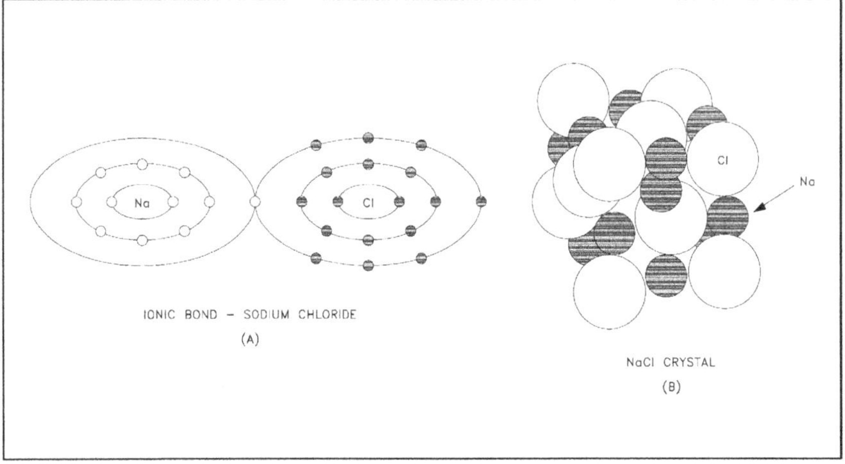

Figure 6 Ionic Bond, Sodium Chloride

The sodium atom loses the one electron in its outer shell to the chlorine atom, which uses the electron to fill its outer shell. When this occurs, the sodium atom is left with a +1 charge and the chlorine atom a -1 charge. The ionic bond is formed as a result of the attraction of the two oppositely-charged particles. No single negatively-charged ion has a greater tendency to bond to a particular positively-charged ion than to any other ion.

Because of this, the positive and negative ions arrange themselves in three dimensions, as shown in Figure 6(B), to balance the charges among several ions. In sodium chloride, for example, each chloride ion is surrounded by as many sodium ions as can easily crowd around it, namely six. Similarly, each sodium ion is surrounded by six chloride ions. Therefore, each chloride ion is bonded to the six nearest sodium ions and bonded to a lesser extent to the more distant sodium ions. Accordingly, the ionic bond is a force holding many atoms or ions together rather than a bond between two individual atoms or ions.

Covalent Bonds

A *covalent bond* is formed when one or more electrons from an atom pair off with one or more electrons from another atom and form overlapping electron shells in which both atoms share the paired electrons. Unlike an ionic bond, a covalent bond holds together specific atoms. Covalent bonding can be single covalent, double covalent, or triple covalent depending on the number of pairs of electrons shared. Figure 7 shows the bonding that occurs in the methane molecule, which consists of four single covalent bonds between one carbon atom and four hydrogen atoms.

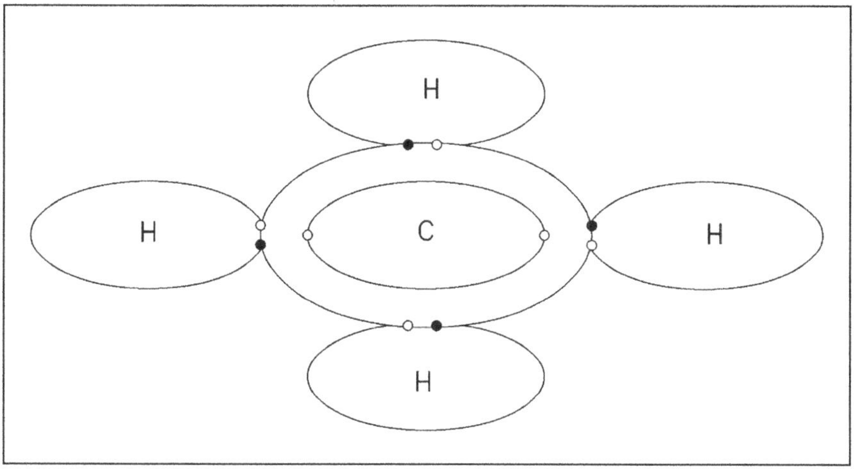

Figure 7 Covalent Bond, Methane CH_4

Two double covalent bonds result when carbon dioxide, which consists of one carbon atom and two oxygen atoms, is formed. Four pairs of electrons are shared by the carbon atom, two from each of the two oxygen atoms as shown in Figure 8. A combination of two electrons form a combination of lower energy than their energy when separated. This energy difference represents the force that binds specific atoms together.

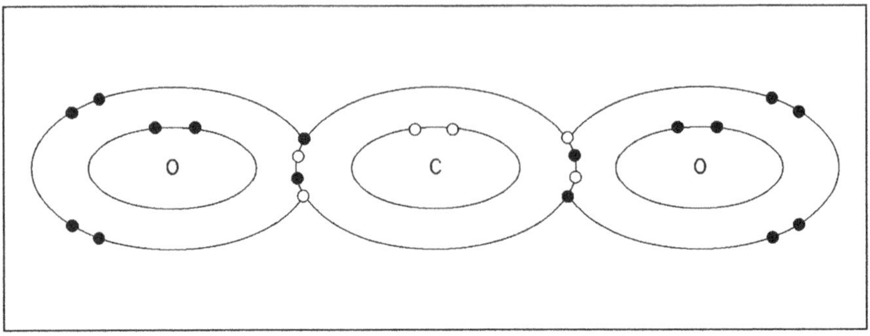

Figure 8 Formation of the Carbon Dioxide Molecule

When both shared electrons in a covalent bond come from the same atom, the bond is called a coordinate covalent bond. Although both shared electrons come from the same atom, a coordinate covalent bond is a single bond similar in properties to a covalent bond. Figure 9 illustrates the bonds of the negatively-charged chlorate ion. The ion consists of one chlorine atom and three oxygen atoms with a net charge of -1, and is formed with two coordinate covalent bonds and one covalent bond. The chlorine atom has effectively gained an electron through the covalent bond, which causes the overall negative charge.

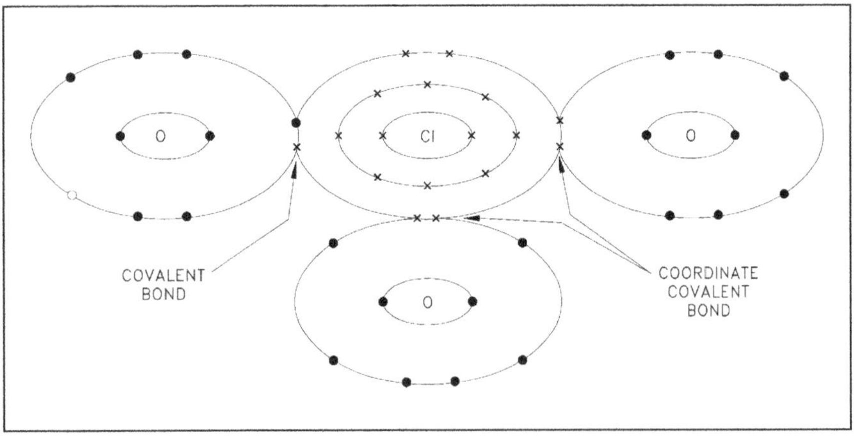

Figure 9 Coordinate Covalent Bond, Chlorate Ion ClO_3

Covalent bonds can be either polar or nonpolar. When the shared pair of electrons is not shared equally, one end of the bond is positive, and the other end is negative. This produces a bond with two poles called a polar covalent bond.

Molecules having polar covalent bonds are called dipolar or *polar molecules*. Water is an example of a polar molecule. When two atoms of the same element share one or more pairs of electrons (such as H or N), each atom exerts the same attraction for the shared electron pair or pairs. When the electron pairs are distributed or shared equally between the two like atoms, the bond is called a nonpolar covalent bond. If all the bonds in a molecule are of this kind, the molecule is called a nonpolar covalent molecule.

Metallic Bonds

Another chemical bonding mechanism is the *metallic bond*. In the metallic bond, an atom achieves a more stable configuration by sharing the electrons in its outer shell with many other atoms. Metallic bonds prevail in elements in which the valence electrons are not tightly bound with the nucleus, namely metals, thus the name metallic bonding. In this type of bond, each atom in a metal crystal contributes all the electrons in its valence shell to all other atoms in the crystal.

Another way of looking at this mechanism is to imagine that the valence electrons are not closely associated with individual atoms, but instead move around amongst the atoms within the crystal. Therefore, the individual atoms can "slip" over one another yet remain firmly held together by the electrostatic forces exerted by the electrons. This is why most metals can be hammered into thin sheets (malleable) or drawn into thin wires (ductile). When an electrical potential difference is applied, the electrons move freely between atoms, and a current flows.

Van der Waals Forces

In addition to chemical bonding between atoms, there is another type of attractive force that exists between atoms, ions, or molecules known as van der Waals forces.

These forces occur between the molecules of nonpolar covalent substances such as H_2, Cl_2, and He. These forces are generally believed to be caused by a temporary dipole, or unequal charge distribution, as electrons constantly move about in an atom, ion, or molecule. At a given instant, more electrons may be in one region than in another region, as illustrated in Figure 10.

The temporary dipole induces a similar temporary dipole on a nearby atom, ion, or molecule. Every instant, billions of these temporary dipoles form, break apart, and reform to act as a weak electrostatic force of attraction known as *van der Waals forces.*

It is important to note that van der Waals forces exist between all kinds of molecules. Some molecules may have these forces, as well as dipole or other intermolecular forces. Van der Waals forces, however, are the only intermolecular bonds between nonpolar covalent molecules such as H_2, Cl_2, and CH_4. The number of electrons in a substance increases as the gram molecular mass (mass in grams of one mole of compound) increases. Therefore, the strength of the van der Waals forces between substances increases with increasing gram molecular mass.

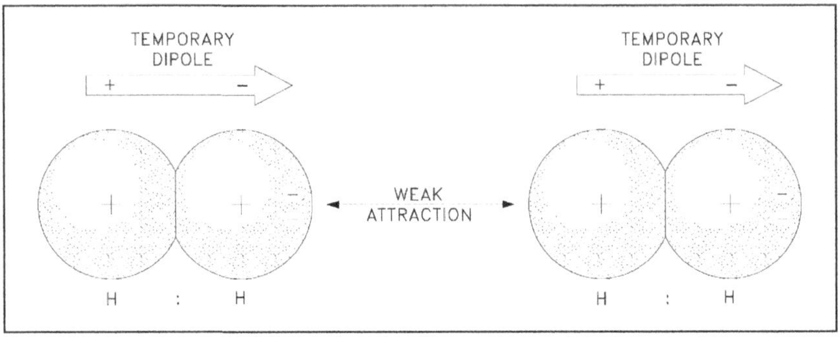

Figure 10 Van der Waals Forces

Van der Waals forces are small compared to the forces of chemical bonding and are significant only when the molecules are very close together.

Organic Chemistry

Organic chemistry is defined as the chemistry of carbon compounds. The compounds of carbon are in forms of living things and in synthetic fabrics and plastics. Organic chemistry is so broad a topic that it is usually subdivided into smaller areas. How the carbon is combined and what it is combined with determines the subdivision for a particular compound. These subdivisions are referred to as families or classes.

The carbon atoms can combine to form straight chains, rings, or branched chains. The bonds between carbon atoms can be single, double, triple or a combination of these. Other atoms (H, O, N, S, P) and the halogens can be attached to the carbon atoms to yield derivatives.

The large family of organic compounds containing only carbon and hydrogen are called hydrocarbons. These can be further divided into two broad classes, aliphatic (fatty) and aromatic (fragrant).

Aliphatic hydrocarbons are divided into two categories, saturated and unsaturated and into subdivisions alkanes, alkenes, and alkynes. These subdivisions reflect the type of bond between the carbon atoms.

Alkanes

Alkanes are saturated compounds which have single bonds between carbon atoms and contain the maximum number of hydrogen atoms possible. Each carbon is flanked by four covalent bonds and each hydrogen atom shares one pair of electrons with a carbon atom, as illustrated in Figure 11.

The general formula for alkanes is C_nH_{2n+2}. The alkanes are colorless, practically odorless, insoluble in water, and readily soluble in nonpolar solvents such as benzene or ether.

Alkanes are low in reactivity. The reactions they do undergo are called halogenation, thermal decomposition (cracking), and combustion. These are summarized below.

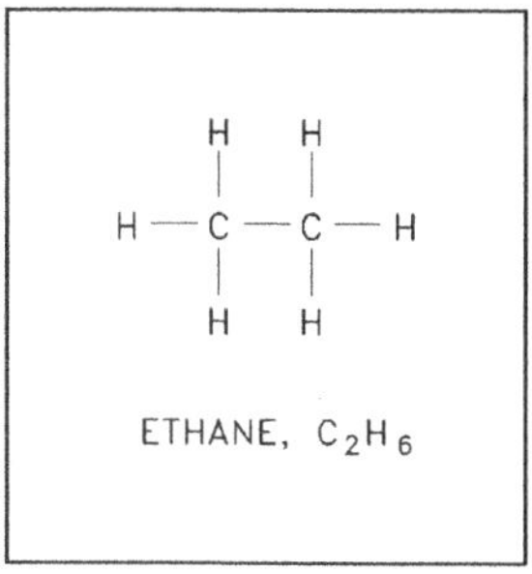

Figure 11 Alkane

Halogenation occurs when a hydrogen atom is replaced with a halogen atom. This is referred to as a substitution reaction. There is no limit to how many hydrogen atoms can be replaced in one molecule.

$$CH_4 + Br_2 \rightarrow CH_3Br + HBr$$

Thermal decomposition or cracking is the process of breaking large molecules into smaller ones. Using heat as a catalyst, propane can be broken into methane and ethylene:

$$CH_4(gas) + 2O_2(gas) \rightarrow CO_2(gas) + 2H_2O(liquid) + 213\ kcal(heat)$$

Combustion occurs when an alkane is burned, the products being carbon dioxide gas, water, and heat. These reactions are highly exothermic and as such the hydrocarbons are frequently used for fuel.

$$C_3H_8 \underset{heat}{\rightarrow} CH_4 + C_2H_4$$

Alkenes

Alkenes are hydrocarbons containing two fewer hydrogen atoms than the corresponding alkane. The general formula for alkenes is C_nH_{2n}. These molecules will have a double bond as illustrated in Figure 12.

Because there are fewer hydrogen atoms than the maximum possible, alkenes are unsaturated hydrocarbons. The chief source for alkenes is the cracking of alkanes.

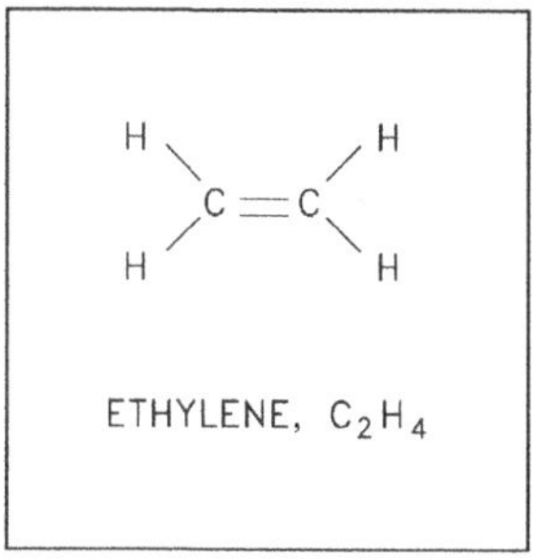

Figure 12 Alkene

$$C_2H_6 \underset{heat}{\rightarrow} C_2H_4 + H_2$$

Alkynes

The third of the aliphatic hydrocarbons are the *alkynes*. These compounds are unsaturated like the alkenes. They contain two fewer hydrogens than the corresponding alkane, C_nH_{2n-2}. The alkyne hydrocarbons contain a triple bond between at least one set of carbon atoms as illustrated in Figure 13.

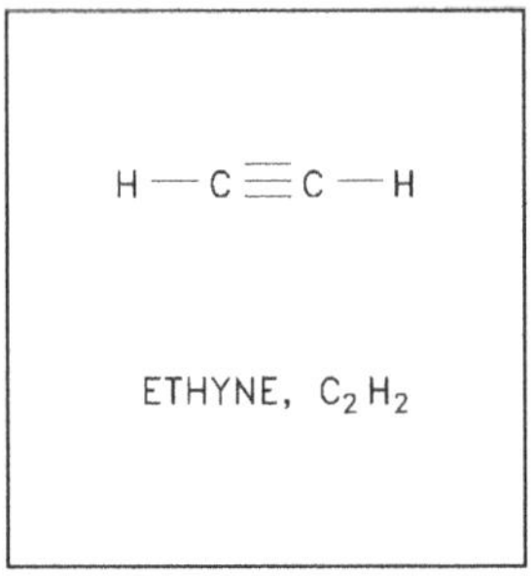

Figure 13 Alkyne

Aromatics

The other broad class of hydrocarbons is the aromatic hydrocarbon. Rather than being arranged in straight chains, as the aliphatics are, these are cyclic formations such as in benzene. The derivatives of cyclic hydrocarbons have pleasant (sometimes toxic) odors. The benzene in rubber cement is a familiar odor. The cyclic compounds have alternating single - double bonds as illustrated in Figure 14.

Aromatic hydrocarbons are very stable chemically, and act very much like alkanes. They will undergo substitution reactions rather than additions.

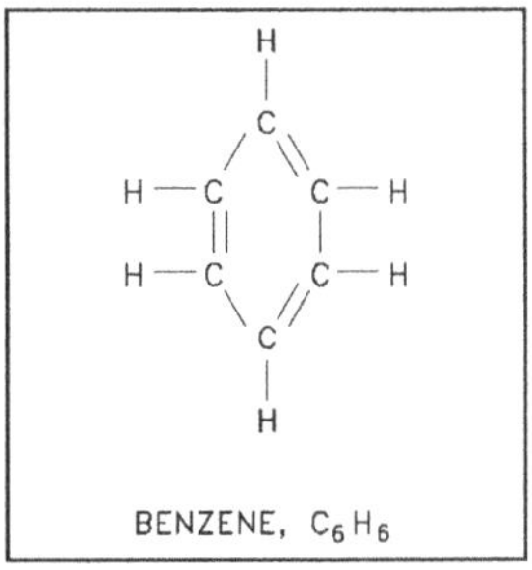

Figure 14 Aromatic

Alcohols

Alcohol is an aliphatic hydrocarbon with a hydroxyl (OH) group substituting for one or more hydrogens as illustrated in Figure 15.

The -OH functional group does not behave in an ionic manner in the case of alcohols. The alcohols are molecular, not ionic, in nature. Alcohols are versatile compounds which are often used to make nearly every other kind of aliphatic compound.

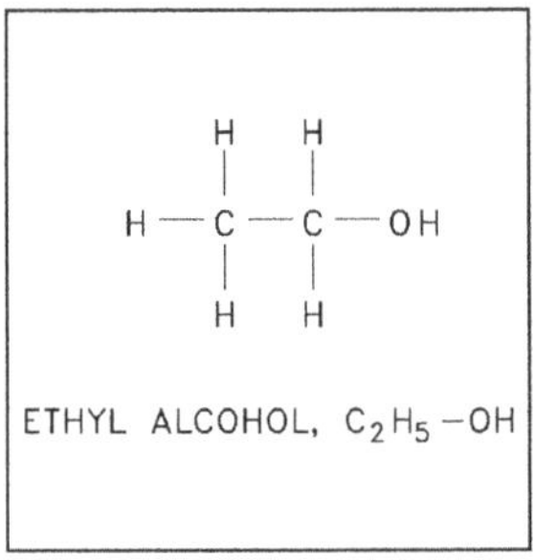

Figure 15 Alcohol

Aldehydes

Aldehydes are one of the oxidation products of the alcohols. Each of these compounds contain a carbonyl group (a carbon atom double bonded to an oxygen atom) as illustrated in Figure 16.

The term "aldehyde" is a contraction of the term "*al*cohol *dehyd*rogenation" indicating that two hydrogen atoms are removed from an end carbon when aldehydes are prepared from primary alcohols. The functional group (-C=O) is always at the end of the carbon chain.

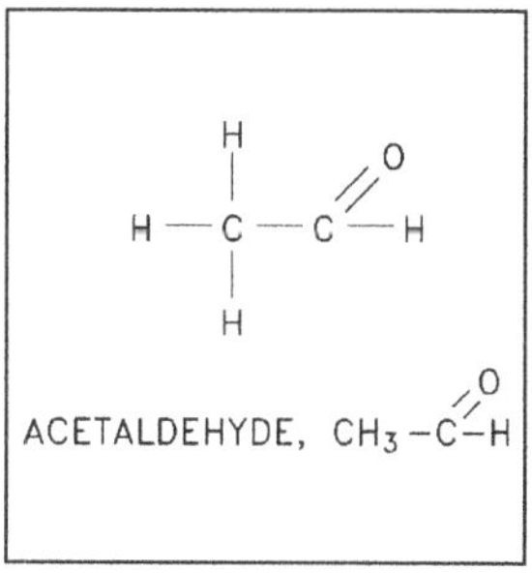

Figure 16 Aldehyde

Basic Chemical Laws

As previously stated, all matter is composed of atoms that which are capable of uniting to form chemical compounds. The various forms of matter can be summarized from a chemical point of view as follows.

1. Molecules are groups or clusters of atoms held together firmly by means of chemical bonding. There are two general types of molecules.

   a. Molecule of an element - Two single atoms of the same element, in certain cases, can become fastened to one another by a chemical bond to form a molecule. Examples of this are hydrogen (H_2), oxygen (O_2), and bromine (Br_2). Most gaseous elements exist as molecules of two atoms.

   b. Molecules of a compound - A compound contains at least two different kinds of atoms. Molecules are the ultimate particles of chemical compounds. Examples of compounds are hydrogen chloride (HCl), water (H_2O), methane (CH_4), and ammonia (NH_3).

2. Elements are substances that cannot be decomposed by ordinary types of chemical change nor made by chemical union.

3. Compounds are substances containing more than one constituent element and having properties different from those of the individual elements. The composition of a specific compound is always definite.

4. Mixtures consist of two or more substances intermingled with no constant percentage composition. Each component retains its original properties.

Forming Chemical Compounds

Chemistry and all other sciences are based on facts established through experimentation. A scientific law is a condensed statement of facts which has been discovered by experiment.

There are three basic laws that apply to chemical reactions. They are the Law of Conservation of Mass, the Law of Definite Proportions, and the Law of Multiple Proportions. These laws are described here to help the reader in understanding the reasons elements and compounds behave as they do.

1. The Law of Conservation of Mass

   This law states that in a chemical reaction the total mass of the products equals the total mass of the reactants. Antoine Lavoisier, a French chemist, discovered that when tin reacts with air in a closed vessel, the weight of the vessel and its contents is the same after the reaction as it was before. Scientists later discovered that whenever energy (heat, light, radiation) is liberated during a reaction, a very small change in mass does occur, but this change is insignificant in ordinary chemical reactions.

2. The Law of Definite Proportions

   This law states that no matter how a given chemical compound is prepared, it always contains the same elements in the same proportions by mass. John Dalton, an English physicist, discovered that when various metals are burned or oxidized in air, they always combine in definite proportions by weight.

   For example, one part by weight of oxygen always combines with 1.52 parts by weight of magnesium or 37.1 parts by weight of tin. This law results from the fact that a compound is formed by the combination of a definite number of atoms of one element with a definite number of atoms of another.

3. The Law of Multiple Proportions

   This law states that if two elements combine to form more than one compound, the masses of one of the elements combining with a fixed mass of the other are in a simple ratio to one another. For example, carbon forms two common compounds with oxygen; carbon monoxide and carbon dioxide. With carbon monoxide (CO), 1.33 grams of oxygen are combined with 1 gram of carbon. With carbon dioxide (CO_2), 2.67 grams of oxygen are combined with 1 gram of carbon. Therefore, the masses of oxygen combining with a fixed mass of carbon are in the ratio 2:1.

Combining Elements

The Laws of Definite Proportions and Multiple Proportions and the related portions of atomic theory form the bases for most quantitative calculations involving chemical reactions. Applying the basic chemical laws to chemical bonding will help the reader to understand the probability and proportions involved in chemical reactions. Regardless of the type of bond (ionic, covalent, coordinate covalent, or metallic), specific amounts of one element will react with specific amounts of the element(s) with which it is combined.

If two substances are placed together in a container, in any ratio, the result is a *mixture*. When a teaspoon of sugar is added to a glass of water, it will slowly dissolve into the water and disappear from view. As a result, the molecules of sugar are evenly distributed throughout the water and become mixed with the water molecules. Because the sugar and water mixture is uniform throughout, it is said to be homogeneous. A homogeneous mixture of two or more substances is called a *solution*. The reason solutions are classified as mixtures rather than as compounds is because the composition is not of fixed proportion.

All solutions consist of a solvent and one or more solutes. The *solvent* is the material that dissolves the other substance(s). It is the dissolving medium. In the water-sugar solution, the water is the solvent. The substances that dissolve in the solution are called *solutes*. In the water-sugar solution, sugar is the solute. It is not always easy to identify which is the solvent and which is the solute (for example, a solution of half water and half alcohol).

Solutions can exist in any of the three states of matter, solid, liquid, or gas. The earth's atmosphere is a gaseous solution of nitrogen, oxygen, and lesser amounts of other gases. Wine (water and alcohol) and beer (water, alcohol, and CO_2) are examples of liquid solutions. Metal alloys are solid solutions (14-karat gold is gold combined with silver or copper).

One factor that determines the degree and/or rate at which a reaction takes place is solubility. *Solubility* is defined as the maximum amount of a substance that can dissolve in a given amount of solvent at a specific temperature. At this point, the solution is said to be saturated. A solution is saturated when equilibrium is established between the solute and the solvent at a particular temperature. *Equilibrium* is the point at which the rates of the forward and reverse reactions are exactly equal for a chemical reaction if the conditions of reaction are constant.

Kinetics is the study of the factors which affect the rates of chemical reactions. There are five principle factors to consider: concentration, temperature, pressure, the nature of the reactants, and the catalyst.

Summary

The important information in this chapter is summarized below.

Chemical Bonding Summary

- The following terms are defined in this chapter:

  An ionic bond occurs where one or more electrons are wholly transferred from one element to another, and the elements are held together by the force of attraction due to the opposite polarity of the charge.

  The van der Waals forces are attractions between the molecules of nonpolar covalent substances. These forces are generally believed to be caused by a temporary dipole, or unequal charge distribution, as electrons constantly move about in an atom, ion, or molecule.

  A covalent bond is a bond in which one or more electrons from an atom pair off with one or more electrons from another atom and form overlapping electron shells in which both atoms share the paired electrons.

  Metallic bonds are bonds where the atoms achieve a more stable configuration by sharing the electrons in their outer shell with many other atoms.

  A mixture is defined as two substances placed together in a container, in any ratio and the composition is not of fixed proportion.

  A solution is a homogeneous mixture of two or more substances.

  A solute is defined as the substance that dissolves in a solution.

  A solvent is defined as the material that dissolves the other substance(s) in a solution. It is the dissolving medium.

  Solubility is defined as the maximum amount of a substance which will dissolve in a given amount of solvent at a specific temperature.

- When the shared pair of electrons which are forming the bond in a molecule are not shared equally, the resulting molecule will have a positive end and a negative end. This type of bond is a polar covalent bond. The molecules are called dipolar or polar molecules.

Chemical Bonding Summary (Cont.)

- In general, the fewer electrons an element must lose, gain, or share to reach a stable shell structure, the more chemically active the element is. The likelihood of elements forming compounds is strongly influenced by the valence shell and on the stability of the resulting molecule. The more stable the molecules are, the more likely these molecules are to form.

- Elements combine in specific ratios, regardless of the type of bond, to form molecules.

- The three basic laws of chemical reactions are discussed:

  (1) The Law of Conservation of Mass - "In a chemical reaction the total mass of the products equal the total mass of the reactants."

  (2) The Law of Definite Proportions - "No matter how a given chemical compound is prepared, it always contains the same elements in the same proportions by mass."

  (3) The Law of Multiple Proportions - "If two elements combine together to form more than one compound, then the masses of one of these elements combining with a fixed mass of the other are in a simple ratio to one another."

CHEMICAL EQUATIONS

This chapter will discuss the method of qualifying solutions. The balancing of the chemical equations will also be explained in this chapter.

EO 2.7 STATE Le Chatelier's principle.

EO 2.8 DEFINE the following terms:

a.	**ppm**	**c.**	**Density**
b.	**Molarity**	**d.**	**Normality**

EO 2.9 BALANCE chemical equations that combine elements and/or compounds.

Le Chatelier's Principle

The effect of temperature on solubility can be explained on the basis of *Le Chatelier's Principle* Le Chatelier's Principle states that if a stress (for example, heat, pressure, concentration of one reactant) is applied to an equilibrium, the system will adjust, if possible, to minimize the effect of the stress. This principle is of value in predicting how much a system will respond to a change in external conditions. Consider the case where the solubility process is endothermic (heat added).

$$\text{solid} + \text{heat} \rightleftharpoons \text{solution}$$

An increase in temperature puts a stress on the equilibrium condition and causes it to shift to the right. The stress is relieved because the dissolving process consumes some of the heat Therefore, the solubility (concentration) increases with an increase in temperature. If the process is exothermic (heat given off).

$$\text{solid} \rightleftharpoons \text{solution} + \text{heat}$$

A temperature rise will decrease the solubility by shifting the equilibrium to the left.

How much solute is dissolved in a solution is very important when the solution is being made for a specific use. To say there is a little, a lot, or a bit would not be very accurate if a specific concentration is required. There are a few common and accurate methods used to express concentration. These are density, molarity, normality, and parts per million.

Density

Density is the measure of the mass per unit volume of a material (density = mass/volume) Density is a characteristic of a substance; mass and volume are not. Mass and volume vary with size but density will remain constant. Temperature will affect the density of a substance and the temperature at which density for that substance was determined is usually reported along with the density value.

Molarity

A useful way to express exact concentrations of solutions is *molarity.* Molarity is defined as moles of solute per liter of solution. Molarity is symbolized by the capital letter M. It can be expressed mathematically as follows.

$$\text{molarity (M)} = \frac{\text{moles of solute (n)}}{\text{liters of solution (V)}}$$

Notice that the moles of solute are divided by the liters of solution not solvent. One liter of one molar solution will consist of one mole of solute plus enough solvent to make a final volume of one liter.

Example 1:

Prepare one molar solution of NaCl.

Solution:

a) Calculate the molecular weight of the salt

1 atom of Na	= 22.989 amu
1 atom of Cl	= 35.453 amu
1 molecule of NaCl	= 58.442 amu

One mole is equal to the gram molecular weight, so one mole = 58.442 grams.

b) 58.442 grams of NaCl is weighed out and sufficient water is added to bring the solution to one liter.

Example 2:

Prepare 3 liters of a 2M NaOH solution.

Solution:

Calculate the amount of NaOH required to prepare the solution.

a) $$\text{molarity (M)} = \frac{\text{moles of solute}}{\text{volume of solution } (\ell)}$$

$$2M = \frac{\text{X moles of solute}}{\text{3 liters}}$$

b) substituting: $$\text{(3 liters) (2M)} = \text{6 moles NaOH}$$

$$\frac{\text{40 grams NaOH}}{\text{1 mole NaOH}} \times \text{6 moles NaOH} = \text{240 grams NaOH}$$

Therefore, to prepare 3 liters of a 2M NaOH solution, 240 grams of NaOH must be weighed out and dissolved in water to make a volume of exactly 3 liters.

Normality

The normal concentration is another method for expressing the concentration of solutions. *Normality* (N) is defined as the number of equivalents of solute dissolved in one liter of solution.

$$\text{Normality (N)} = \frac{\text{equivalents of solute}}{\text{liter of solution}}$$

One equivalent of acid is the amount of acid necessary to give up one mole of hydrogen ions in a chemical reaction. One equivalent of base is the amount of base that reacts with one mole of hydrogen ions. When expressing the concentrations of bases, normality refers to the number of available hydroxyl ions. Because hydrogen and hydroxyl ions combine on a one-to-one basis, one OH^- is equivalent to one H^+ ion.

a) H^+Cl^- + Na^+OH^- → Na^+Cl^- + H^+OH^-
Acid Base Salt Water

b) H_2SO_4 + $2K^+OH^-$ → K_2SO_4 + $2H^+OH^-$
Acid Base Salt Water

Notice that in reaction a), one mole of HCl yields one equivalent per mole or one mole of H^+ ions. H_2SO_4 has two equivalents per mole or two H^+ ions because each mole of the compound can release two moles of hydrogen ions.

The number of equivalents of an acid or base can be determined from equivalent weight. The *equivalent weight* is defined as the molecular weight of the acid or base divided by the number of replaceable hydrogen or hydroxyl ions.

Example:

The equivalent weight of H_2SO_4 is:

$$\frac{98.0734 \text{ grams (molecular wt.)}}{2 \text{ H}^+ \text{ ions}} = 49 \text{ grams/equivalent.}$$

This means 49 grams of H_2SO_4 is equal to one equivalent of H_2SO_4. If there is a one liter solution that contains 78.32 grams H_2SO_4, the number of equivalents is:

$$\frac{78.32 \text{ grams } H_2SO_4}{1} \times \frac{1 \text{ equivalent}}{49 \text{ grams } H_2SO_4} = 1.6 \text{ equivalents.}$$

Since normality is equal to the number of equivalents per liter, the normality of this solution is 1.6 equivalents/liter, or 1.6 N.

Parts per Million

Another term used to describe the specific concentration of a solution is parts per million or ppm. The term *ppm* is defined as the concentration of a solution in units of one part of solute to one million parts solvent. One ppm equals one milligram of solute per liter of solution. Another term, parts per billion (ppb), is defined as one part solute per one billion parts solvent. One ppb is equal to one microgram solute per liter of solution. These two terms are usually used for very dilute solutions.

Chemical Equations

The chemical equation is simply a representation of a chemical reaction in the chemist's shorthand. In a chemical equation, the substances originally present are called the reactants and the new substances being formed are called the products. In the chemical reaction, the reactants are on the left of the arrow and the products are on the right of the arrow. For example, when looking at the combination of hydrogen and oxygen into water, the water molecule would be the product on the right.

$$2H_2 + O_2 \rightarrow 2H_2O$$

If looking at the decomposition of water, the hydrogen and oxygen would be the products and the water is the reactant. The arrow means yields and is used to separate the reactants from the products. It also indicates the direction of the reaction as illustrated below.

$$2H_2O \rightarrow 2H_2 + O_2$$

The single arrow in the above equation implies that the reaction will proceed in only one direction until it subsides or stops. If the reaction is reversible, that is, it can proceed left to right and right to left until it reaches an equilibrium, then a double arrow is used.

$$2H_2O \leftrightarrows 2H_2 + O_2$$

When writing an equation, always place the reactant on the left and the products on the right even in the case of a reversible reaction.

A chemical equation represents not only the reaction, but also shows the number of atoms or molecules entering into and produced by the reaction. The formulas must be balanced correctly based on the valences of the constituent elements.

Balancing Chemical Equations

The number of atoms or molecules of each substance is shown by the coefficients in the equation. Because atoms cannot be created or destroyed in a chemical reaction, a chemical equation must be balanced so that there are exactly the same number of atoms of each element on each side of the equation.

Example:

Explain the following chemical equation.

$$2Fe + 3H_2O \rightarrow Fe_2O_3 + 3H_2\uparrow$$

Solution:

This chemical equation shows that iron reacts with water to form ferric oxide and hydrogen gas (the vertical arrow ↑ indicates a gas). This chemical equation also shows that for every two atoms of iron that react, three molecules of water are used to form one molecule of ferric oxide and three molecules of hydrogen gas. This is a balanced chemical equation. There are two iron atoms on each side of the equation; there are six hydrogen atoms on each side; and there are three oxygen atoms on each side.

There are no fixed rules for balancing chemical equations. Learning how is a matter of practice. The balancing of most equations can be accomplished by following the guidelines explained below.

Guidelines:

a. Once the correct chemical formula for a compound is written in an equation, do not modify it.

b. Select the compound with the greatest number of atoms. Then begin by balancing the element in that compound with the most atoms. There must be the same number of atoms of an element on each side of the equation. As a rule of thumb, this first element should not be hydrogen, oxygen, or a polyatomic ion.

c. Balance the atoms of each element in the compound by placing the appropriate coefficient in front of the chemical symbol or formula.

d. Next, balance the polyatomic ions. In some cases, the coefficient assigned in guideline 2 may have to be changed to balance the polyatomic ion.

e. Balance the hydrogen atoms next, then the oxygen atoms. If these elements appear in the polyatomic ion it should not be necessary to balance them again.

f. All coefficients will be whole numbers. The coefficients should be reduced to the lowest possible ratios.

g. As simple as it sounds, check off each element as it is accounted for since this will prevent double inclusion or a missed atom.

Example 1:

$$FeS_2 + O_2 \rightarrow Fe_2O_3 + SO_2$$

Solution 1:

Starting with Fe_2O_3 (see guideline b), write $2FeS_2 + O_2 \rightarrow Fe_2O_3 + SO_2$, which balances the Fe on each side of the equation.

Now there are 4 S atoms on the left side so balance the S atoms by writing
$2FeS_2 + O_2 \rightarrow Fe_2O_3 + 4SO_2$.

Everything is balanced except the O. There are now 2 O atoms on the left and 11 O atoms on the right. To get 11 O atoms on the left write
$2FeS_2 + 5.5O_2 \rightarrow Fe_2O_3 + 4SO_2$.

This makes both sides of the equation balanced except the coefficients must be whole numbers (guideline f). To meet guideline f, multiply both sides by two which will bring the left side to a whole number of O_2 molecules.

Thus, the solution is $4FeS_2 + 11O_2 \rightarrow 2Fe_2O_3 + 8SO_2$.

Example 2:

$$NH_3 + CuO \rightarrow H_2O + N_2 + Cu$$

Solution 2:

Start with NH_3 since there are two N atoms on the right of the equation. To balance the N atoms write $2NH_3 + CuO \rightarrow H_2O + N_2 + Cu$ (guideline b).

Since the H appears in only the NH_3 and H_2O and the NH_3 has been balanced, the H_2O will be balanced. Write $2NH_3 + CuO \rightarrow 3H_2O + N_2 + Cu$ (guideline e).

Oxygen appears only in CuO and in H_2O, and since the H_2O has been already been balanced write $2NH_3 + 3CuO \rightarrow 3H_2O + N_2 + Cu$ (guideline e).

That leaves the Cu to be balanced. Thus, the solution is
$2NH_3 + 3CuO \rightarrow 3H_2O + N_2 + 3Cu$.

Example 3:

$Na_2CO_3 + Ca(OH)_2 \rightarrow NaOH + CaCO_3$

Solution 3:

There are two Na atoms on the left so start with the Na by writing $Na_2CO_3 + Ca(OH)_2 \rightarrow 2NaOH + CaCO_3$ (guideline b).

By adding the 2, the equation is now completely balanced. This equation illustrates that not all equations are that hard to balance.

Most chemical equations do not indicate a number of important facts about the chemical reactions they represent. Chemical equations do not necessarily describe the path by which the substances reacting are converted to products.

$$2H_2 + O_2 \rightarrow 2H_2O$$

The equation would seem to imply that two molecules of hydrogen collide with one molecule of oxygen, and two molecules of water are produced. The actual mechanism by which this reaction takes place is much more complicated and involves a series of processes. Chemical equations do not indicate the rate at which the reaction proceeds, or even whether the reaction will occur in a finite time. In many cases, reactions will occur only under a particular set of circumstances and then only at a definite rate. Chemical equations do not show whether the reaction proceeds to completion or, if incomplete, the extent of reaction. In most cases, the substances that react never completely disappear; however, their concentration may be exceedingly small. Reactions that do not go to completion are usually represented in chemical equations by using double horizontal arrows $\leftrightarrows$. In general, a reaction will go to completion only if one or more of the products is removed from the field of the reaction. This is often accomplished if one of the products is a gas or is insoluble in the reaction mixture.

In the discussion of chemical equations, emphasis is normally placed on the number of atoms or molecules involved in the reaction. However, chemical equations are very effective in representing chemical reactions on a macroscopic scale. Practical chemical calculations involve very large numbers of atoms and molecules.

The equation weight in grams of a compound or element is defined as the gram molecular weight times the number of molecules of the compound, as shown by the coefficients of the chemical equation for the reaction. The sums of the equation weights on each side of a chemical equation must be equal. Chemical calculations are based on the fact that every fraction or multiple of the equation weights of substances that react gives a corresponding fraction or multiple of the equation weights of the products of the reaction. In other words, if 30 grams of a substance that has an equation weight of 15 grams reacts with some amount of another substance to form a product with an equation weight of 20 grams, then 40 grams of that product will be formed.

Example:

How many grams of ferric oxide will be formed if 27.9 grams of iron reacts with water according to the following equation.

$$2Fe + 3H_2O \rightarrow Fe_2O_3 + 3H_2\uparrow$$

Solution:

The equation weight of iron equals the gram atomic weight of iron times the number of atoms shown reacting in the equation, which is two. Using Table 2:

$$\begin{aligned} \text{Equation Weight Fe} &= 2 \times 55.8 \text{ grams} \\ &= 111.6 \text{ grams} \end{aligned}$$

Because 27.9 g of iron react, the fraction of the equation weight that reacts is:

$$\frac{27.9 \text{ grams}}{111.6 \text{ grams}} = \frac{1}{4}.$$

Thus, 1/4 of the equation weight of ferric oxide will be formed.

The equation weight of ferric oxide equals the gram molecular weight of ferric oxide times the number of molecules shown formed in the equation, which is one.
Using Table 2:

$$\begin{aligned} \text{Equation Weight } Fe_2O_3 &= 2(55.8 \text{ g}) + 3(16.0 \text{ g}) \\ &= 111.6 \text{ g} + 48.0 \text{ g} \\ &= 159.6 \text{ g} \end{aligned}$$

Thus, the amount of ferric oxide formed is:

$$\frac{1}{4}(159.6 \text{ g}) = 39.9 \text{ g}.$$

Summary

The important information from this chapter is summarized below.

Chemical Equations Summary

- Le Chatelier's principle states that if a stress is applied to an equilibrium, the system will adjust to minimize the effect of the stress.

- The following terms are defined:

  Density is the measure of the mass per unit volume of a material (density = mass/volume).

  Molarity is moles of solute per liter of solution. Molar solutions are symbolized by the capital letter M.

  Normality, (N), is defined as the number of equivalents of solute dissolved in one liter of solution.

  ppm expresses the concentration of a solution in units of one part of solute to one million parts solvent. One ppm equals one milligram of solute per liter of solution.

- To balance chemical equations, the number of elemental constituents must be the same on each side of the equation.

ACIDS, BASES, SALTS, AND pH

Different substances respond differently in solution. How substances behave in water is of special interest to the power industry. The interactions of water and acids, bases, or salts are of special interest because water is used in many industries. This chapter will introduce the student to the general behavior of these substances.

EO 3.1 DEFINE the following terms:

a.	**Acid**	**e.**	**Base**
b.	**Salt**	**f.**	**pH**
c.	**pOH**	**g.**	**Dissociation constant of water**
d.	**Alkalies**		

EO 3.2 STATE the formula for pH.

EO 3.3 STATE the formula for pOH.

EO 3.4 CALCULATE the pH of a specified solution.

Acids

Substances that form ions when they are dissolved in water are called electrolytes. The three types of electrolytes are acids, bases, and salts.

Acids are substances that dissociate in water to produce hydrogen (H^+). An example of a common acid is sulfuric acid, H_2SO_4. In solution, H_2SO_4 dissociates to form hydrogen and sulfate ions according to the following equation. The designation in parenthesis in the following formulas indicate the state of the reactants and the products, i.e., liquid (aq). solid (s), or gas (g).

$$H_2SO_{4(aq)} \rightarrow 2H^+ + SO_4^{--}$$

Additional examples of acids are vinegar, aspirin, and lemon juice. These substances share the following common properties.

1. Acid solutions taste sour (acid means "sour" in Latin).

2. Acids react with many metals to form hydrogen gas.

$$\underset{\text{acid}}{2HCl_{(aq)}} + \underset{\text{metal}}{Zn_{(s)}} \rightarrow \underset{\text{metal salt}}{ZnCl_{2(aq)}} + \underset{\text{hydrogen gas}}{H_{2(g)}}$$

3. Acids turn litmus paper red.

4. Acid solutions conduct electricity.

5. Acids react with bases to form a salt and water.

   Example:

$$\underset{\text{Acid}}{HNO_{3(aq)}} + \underset{\text{Base}}{KOH} \rightarrow \underset{\text{Salt}}{KNO_3} + \underset{\text{Water}}{H_2O}$$

6. Acids react with carbonates to form CO_2 (carbon dioxide gas).

   Example:

$$\underset{\text{Acid}}{2HCl_{(aq)}} + \underset{\text{Carbonate}}{CaCO_3} \rightarrow CaCl_2 + CO_2\uparrow + H_2O$$

Bases

Bases are substances that produce hydroxide ions (OH^-) in water solutions. An example of a common base is sodium hydroxide, NaOH. In solution, it dissociates to form sodium ions and hydroxide ions according to the following equation:

$$\underset{\text{Base}}{NaOH_{(aq)}} \rightarrow Na^{+}_{(aq)} + OH^{-}_{(aq)}.$$

Common types of bases are lye, household ammonia, and most soaps. The following are four characteristic properties of bases.

1. Basic solutions taste bitter and feel slippery to the touch.

2. Bases turn litmus paper blue.

3. Basic solutions conduct electricity.

4. Bases neutralize acids.

   Example:

$$\underset{\text{Base}}{2NaOH} + \underset{\text{Acid}}{H_2SO_4} \rightarrow \underset{\text{Salt}}{Na_2SO_4} + \underset{\text{Water}}{2H_2O}$$

Salts

When an acid reacts with a base, two products are formed; water and a salt. A *salt* is an ionic compound composed of positive ions and negative ions. The ionic bond is what keeps salts in their molecular form. Some compounds look like salts, but are actually covalent compounds (have a covalent bond).

Example:

$$\underset{\text{Acid}}{HCl_{(aq)}} + \underset{\text{Base}}{NaOH_{(aq)}} \rightarrow \underset{\text{Salt}}{Na\,Cl} + \underset{\text{Water}}{H_2O}$$

This type of acid-base reaction is called neutralization.

Certain soluble salts (principally sodium, potassium, magnesium, and calcium) that have the property of combining with acids to form neutral salts are called *alkalies*. Two of the common salts are sodium chloride (table salt) and calcium chloride (salt put on the road in winter). Unlike acids and bases, salts vary greatly in all their properties except their ionic characteristics. Salts may taste salty, sour, bitter, astringent, sweet, or tasteless. Solutions of salts may be acidic, basic, or neutral to acid-base indicators. Fused salts and aqueous solutions of salts conduct an electric current. The reactions of salts are numerous and varied.

pH

Many compounds dissolve in water and alter the hydrogen ion concentration. Compounds that produce hydrogen ions directly when dissolved in water are called acids, and those that produce hydroxyl ions directly when dissolved in water are called bases. To treat these aspects of chemistry more precisely, a quantitative system of expressing acidity or basicity (alkalinity) is needed. This need could be met by using the value of $[H^+]$, where $[H^+]$ is expressed in moles/liter, as a measure of acidity. But, in most cases $[H^+]$ is in the range of 10 to 10^{-14} moles/liter. Because numbers of this magnitude are inconvenient to work with, an alternate system for expressing the acidity of dilute solutions has been devised. This system is based on a quantity called *pH*. ph is defined as the negative logarithm of the hydrogen concentration, represented as $[H^+]$ in moles/liter.

$$pH = -\log [H^+] \quad (1\text{-}1)$$

$$[H^+] = 10^{-pH} \quad (1\text{-}2)$$

The negative logarithm is specified because the logarithm of any number less than 1 is negative; thus multiplication by -1 causes the values of pH to be positive over the range in which we are interested. (The term pH was first defined by a Danish chemist and is derived from p for the Danish word potenz (power) and H for hydrogen.)

Example 1: The hydrogen concentration, $[H^+]$, of a solution is 4.8×10^{-8} moles/liter. What is the pH of the solution?

Solution 1:

$$pH = -\log [H^+]$$
$$= -\log (4.8 \times 10^{-8})$$
$$= 7.32$$

Example 2: The pH of a solution is 3.83. What is the hydrogen concentration of the solution?

Solution 2:

$$[H^+] = 10^{-pH}$$
$$= 10^{-3.83} \text{ moles/liter}$$
$$= 1.48 \times 10^{-4} \text{ moles/liter}$$

pOH

It is also important to understand the relationship between the $[H^+]$ and the $[OH^-]$ concentrations. The *pOH* of a solution is defined as the negative logarithm of the hydroxyl concentration, represented as $[OH^-]$ in moles/liter.

$$pOH = -\log [OH^-] \quad (1\text{-}3)$$

$$[OH^-] = 10^{-pOH} \quad (1\text{-}4)$$

For water solutions, the product of the hydrogen ion concentration and the hydroxyl concentration is always 1×10^{-14} at 25°C. This means that the sum of pH and pOH is equal to 14 under these conditions.

$$[H^+] \times [OH^-] = 1 \times 10^{-14}$$

$$pH + pOH = 14$$

The following example illustrates this point.

Example: What is the hydrogen ion concentration $[H^+]$ and the hydroxyl concentration $[OH^-]$ in a solution with a pH of 5.5?

Solution:

$$[H^+] = 10^{-pH}$$
$$= 10^{-5.5}$$
$$= (10^{-6})\,(10^{.5})$$
$$= 3.16 \times 10^{-6} \text{ moles/liter}$$

since

$$pOH + pH = 14$$
$$pOH = 14 - 5.5$$
$$pOH = 8.5$$

and

$$[OH^-] = 10^{-pOH}$$
$$= 10^{-8.5}$$
$$= (10^{-9})\,(10^{.5})$$
$$= 3.16 \times 10^{-9} \text{ moles/liter}$$

Dissociation Constant

The product of ionic concentrations, $K_w = [H^+]\,[OH^-]$, is called the Ion Product Constant for water, or more frequently, the Ionization Constant or *Dissociation Constant*. At 25°C, K_w equals 1×10^{-14}. K_w varies with temperature and, at 37°C (body temperature), the value is about 3.4×10^{-14}. Figure 17 illustrates the relationship of the Dissociation Constant with water temperature. As indicated earlier, a change in temperature causes a change in equilibrium in a chemical reaction. Because the ionization of water is an equilibrium process, it is also affected by temperature. The Dissociation (or Ionization) Constant, K_w, for water increases with increasing temperature up to about 500°F and then decreases. Experimental values of K_w at various temperatures are listed in Table 5 and graphed in Figure 16. Notice in Table 5 that the pH of pure water changes with temperature. For pure water at any temperature, however, $[H^+] = [OH^-]$. It should be noted that the equation pH + pOH = 14 is true only at or near 25°C (77°F).

Example: The hydroxyl concentration $[OH^-]$ of a water solution° at 25 C is 7.2×10^{-9} moles/liter. Calculate the pH of the solution.

Solution:

$$K_w = [H^+]\,[OH^-]$$

$$[H^+] = \frac{K_w}{[OH^-]}$$

$$= \frac{1 \times 10^{14}}{7.2 \times 10^{-9}}$$

$$= 1.38 \times 10^{6}$$

$$pH = -\log\,[H^+]$$

$$= -\log\,(1.38 \times 10^{6})$$

$$= 5.86$$

TABLE 5
Ion Product Constant and Neutral pH for Water at Various Temperatures

Temperature		Ion Product Constant for Water	pH of Pure Water
F	C		
64.4	18	0.64×10^{-14}	7.10
77	25	1.0×10^{-14}	7.00
140	60	8.9×10^{-14}	6.54
212	100	6.1×10^{-13}	6.10
302	150	2.2×10^{-12}	5.83
392	200	5.0×10^{-12}	5.65
482	250	6.6×10^{-12}	5.59
572	300	6.4×10^{-12}	5.60
662	350	4.7×10^{-12}	5.66

At 25°C, an equilibrium exists between pure molecular water and its ions. The $[H^+]$ equals the $[OH^-]$ and both have values of 1×10^{-7} moles/liter. Using the pH definition, it follows that the pH of pure water at 25°C is 7. pH values less than 7 indicate an acidic solution and values greater than 7 indicate a basic or alkaline solution.

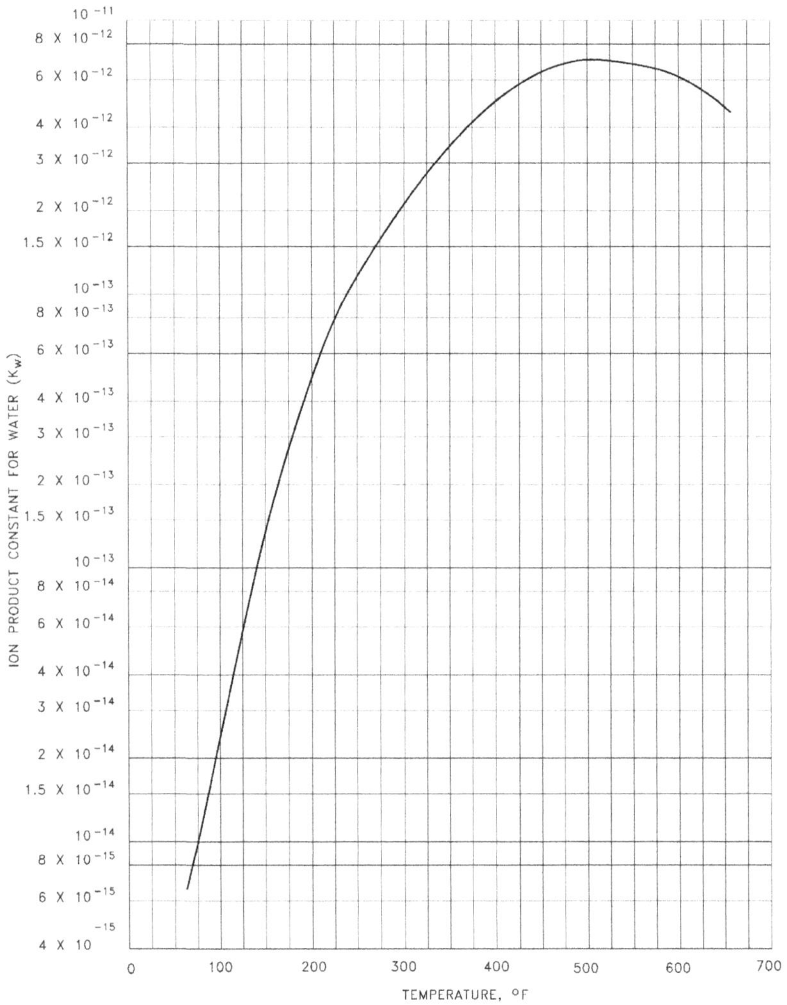

Figure 17 Ion Product Constant for Water

Summary

The important information in this chapter is summarized below.

Acids, Bases, Salts, and pH Summary

- The following terms are defined in this chapter:

  Acid - substances that produce hydrogen (H^+) in water solutions

  Base - substances that produce hydroxide ions (OH^-) in water solutions

  Salt - a compound composed of positive and negative ions held together with an ionic bond

  Alkalies - certain soluble salts, principally sodium, potassium, magnesium and calcium, that have the property of combining with acids to form neutral salts

  pH - the negative logarithm of the hydrogen concentration $[H^+]$ in moles/liter

  pOH - the negative logarithm of the hydroxyl concentration $[OH^-]$ in moles/liter

  Dissociation constant of water (K_w) - the product of ionic concentrations, ($[H^+]$ $[OH^-]$); at 25°C $K_w = 1 \times 10^{-14}$

- The formula for pH is: $pH = -\log [H^+]$
  Therefore, $[H^+] = 10^{-pH}$

- The formula for pOH is: $pOH = -\log [OH^-]$
  Therefore, $[OH] = 10^{-pOH}$

Intentionally Left Blank.

Department of Energy
Fundamentals Handbook

PLANT CHEMISTRY

Appendix A
Basic Separation Theory

BASIC SEPARATION THEORY

Introduction

The fundamental principles upon which a gaseous diffusion plant is based are as follows.

1. All the particles (atoms, molecules, or ions) which make up the gas are continuously moving in straight lines in all directions. The particles collide with anything in their path (e.g., other particles or the walls of the container), thereby altering the course of moving particles but not their average velocities.

2. All of the particles have the same average kinetic energy. Accordingly, if the masses of the particles are different, so must their velocities be different with the lighter particles having the greater average velocities.

On the basis of the two principles stated above, if a quantity of gas is confined by a porous membrane or barrier, some of the gas will escape through the pores in the barrier. If the confined gas is isotopic (a mixture of particles of different molecular weights), the lighter particles of the gas will have a greater tendency to diffuse through the barrier because of their greater average velocities. Consequently, the gas which has passed through the barrier will be enriched in the light constituent while that gas which has not diffused through the barrier will be depleted in the light constituent.

Isotopic Separation

The isotopes with which we are concerned are those of uranium; namely U^{235} and U^{238}. In order that the isotopes of uranium may be separated by the gaseous diffusion process, it is necessary that the uranium be in a gaseous state. This is accomplished by combining the uranium with fluorine to form uranium hexafluoride, which is a gas at workable temperatures and pressures. Another advantageous quality of uranium hexafluoride is that fluorine has no isotopes to further complicate the separation process. In general, the following discussion is concerned with the separation of $U^{235}F_6$ and $U^{238}F_6$.

Separation Factor

The ability of a barrier to separate isotopes is expressed by a value called the *separation factor*. It is denoted by the Greek letter α, and is defined as the mole ratio of light component to heavy component in the gas passing through the barrier at a given point divided by a similar mole ratio on the high pressure surface of the barrier at the same point. The separation factor can be expressed by the following equation.

$$\alpha = \frac{\dfrac{y}{1-y}}{\dfrac{x}{1-x}}$$

In this equation, y and x are the mol fraction of light component on the low and high pressure sides of the barrier respectively. It should be emphasized that α, as shown in Figure A-1, is a point value, and that x and y are concentrations at a given point on the two surfaces of the barrier. In practice, point values are usually not determined. The average α's of the surface of samples are calculated from experimental separation data.

As previously stated, the separation of isotopes is dependent upon the difference in the mass of the particles. Under ideal conditions, the separation factor would be equal to the ratio of the square roots of the molecular weights. For a mixture of $U^{235}F_6$ and $U^{238}F_6$, this value is 1.0043 and is commonly referred to as the ideal separation factor. In actual practice, however, these ideal conditions cannot be duplicated. For example, to approach the ideal separation factor, only a small fraction of the total gas involved could be diffused through the barrier. For such a process to be productive, the plant size and its power consumption would be too large to be economically feasible. Limitations of plant size and power consumption have, necessitated a reduction in the separation factor to a lower value. This lower value is called the actual separation factor.

Stage Separation

The ratio of the total molar rate of flow through the barrier within a converter to the total rate of flow into the same converter is called the *cut* and is designated by the Greek letter Θ.

$$\Theta = \frac{\textit{Molar flow of enriched stream}}{\textit{Molar flow of input stream}}$$

In a production cascade, a cut of approximately 0.5 is the optimum value with respect to operating efficiency and production.

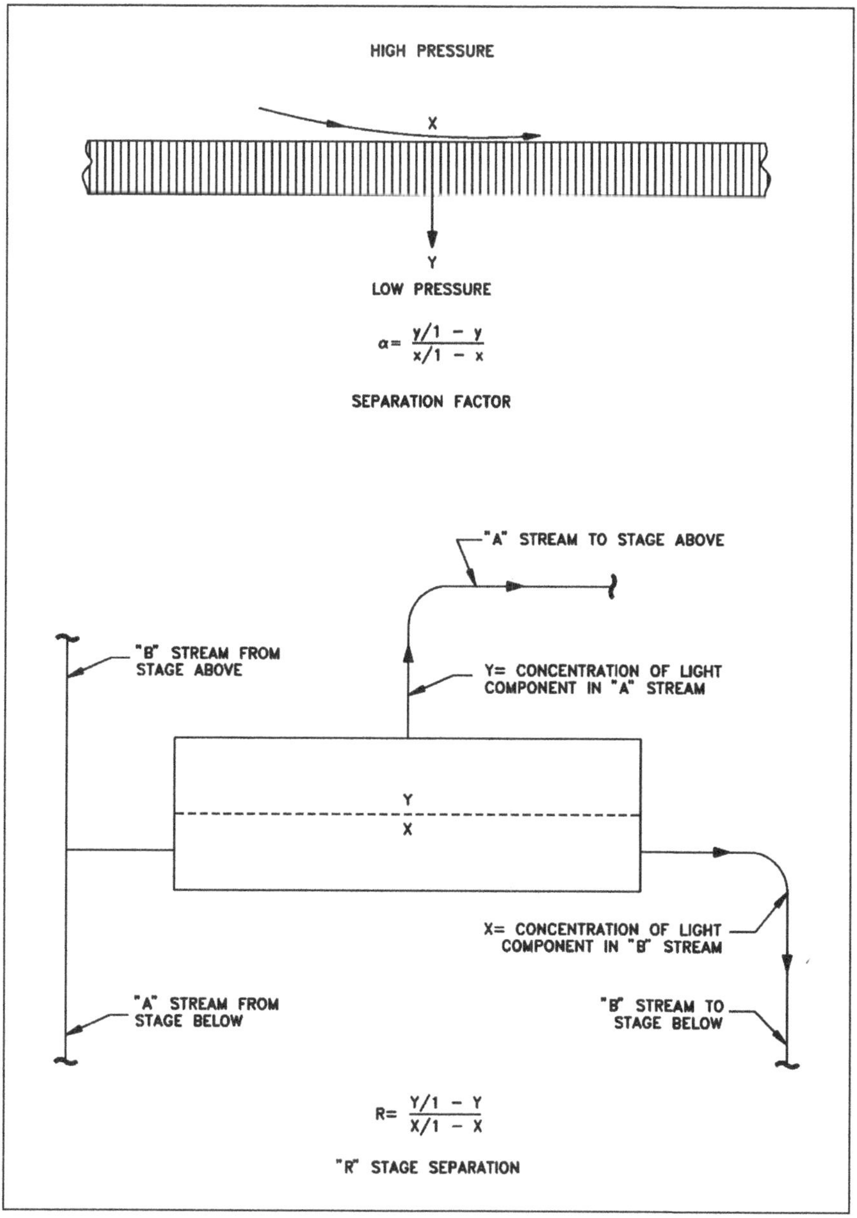

Figure A-1 "R" Stage Separation

The actual separation achieved by a stage is expressed in terms of the mol fraction of the light component and for the purpose of this discussion is designated by the capital letter R. It can be defined as follows: if Y is the concentration of light component in the enriched stream leaving a converter and X is its concentration in the depleted stream leaving the same converter, then

$$R = \frac{\dfrac{Y}{1 - Y}}{\dfrac{X}{1 - X}}.$$

This definition is similar to that of the separation factor, but, as shown in Figure A-1, the value of R is based on the overall separation of a stage while that of the separation factor is based on the performance of the barrier alone. The relation between the two is given by the following equation.

$$R = E \alpha \frac{1}{\Theta} \ln \frac{1}{1 - \Theta}$$

E is a correction factor called the *mixing efficiency*. It can be explained as follows: since the light component passes through the barrier more rapidly than the heavy component, the layer of gas at the surface on the high pressure side of the barrier becomes more depleted in the light component than the bulk of the gas on the high pressure side. Thus, there is a concentration gradient perpendicular to the surface of the barrier which results in a decrease in separation. In the particular process of separation of uranium isotopes, the value of R in the preceding equation differs very little from unity ($R = 1.003$) and is somewhat inconvenient for use in calculations. By subtracting 1 from each side of the equation, equating the symbol ψ (psi) to $R - 1$, and by making a slight approximation, the equation becomes

$$\psi = \frac{Y - X}{X(1 - X)}.$$

This equation is used frequently in the plant in making separation tests.

Barrier Measurements

The ability of a barrier to transmit gas is usually expressed as γ (gamma), the *permeability*. This is a measure of the total flow through a barrier and may be defined as the ratio of the number of molecules which pass through the barrier to the number which would pass through the space if the barrier were not there. This total flow through a barrier is known to be a combination of several different types; namely, diffusive flow, film flow, and viscous flow. As was previously explained, the type of flow that is of greatest importance for a gaseous diffusion plant is diffusive flow. However, as film flow and viscous flow affect the separation through a barrier, they cannot be ignored.

Film flow refers to the transport of molecules under the influence of the force field between the molecules and the barrier surface. When such force fields are significant, a large number of molecules will not have sufficient velocity normal to the surface to escape from this field so that their motion will be confined to the barrier surface and barrier pore surface. Such molecules would pass through the barrier as a film on the surface of the pores.

Viscous flow occurs when molecules flow as a group in the manner of ordinary flow through a tube. Referring back to permeability, it is known that the flow through a barrier increases as the pressure increases. Also, as the pressure increases, so does the viscous component of flow. The relation between permeability and viscous flow can best be shown by the following equation.

$$\gamma = \gamma_o \left[1 + S(P_f + P_b)\right]$$

In this equation where P_f equals the fore or high side pressure and P_b equals the back or low side pressure. γ_o equals the permeability, γ, when

$$P_f + P_b = 0.$$

S is called the slope factor and is inversely proportional to the viscosity of the gas.

In the above equation, the term $S(P_f + P_b)$ is a measure of the viscous component of flow. Figure A-2 illustrates the variation of permeability with the slope factor and changes in pressure.

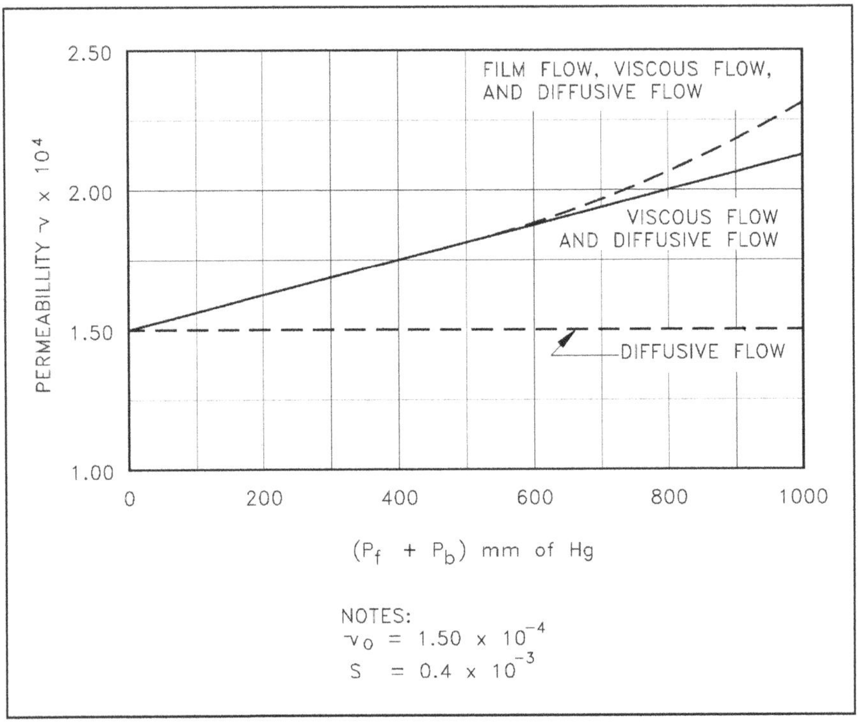

Figure A-2
Variation of Permeability with the Slope Factor and Change in Pressure

It should be noted that in practice, permeability is expressed in two different ways; design and actual. Design permeability is expressed as a ratio as previously stated. It is usually determined by testing the barrier with a non-toxic gas such as nitrogen. Actual permeability is expressed as a percentage of the design permeability and is determined from tests with the barrier in actual operation in the cascade. Its primary value is in determining the performance of a barrier after being in use for a period of time.

Cascade Theory

In the operation of a diffusion cascade, it is often necessary to charge the plant with feed material of different assays or concentrations of $U^{235}F_6$. This feed must be fed into the cascade at a point where the assay of the gas already being processed is the same as that of the feed to be charged.

To choose this feed point, therefore, it must be possible to calculate the concentration of $U^{235}F_6$ in the process gas at any point in the cascade. These calculations can be made by knowing two things: first, the concentration of $U^{235}F_6$ in the process gas at several points in the cascade (determined by laboratory and instrument analysis); and second, the increase in the concentration of $U^{235}F_6$ accomplished by each stage. This is called the stage enrichment and should not be confused with the separation accomplished by a stage previously discussed. Therefore, by knowing the stage enrichment and the $U^{235}F_6$ concentration at some point in the cascade, it is possible to determine the concentration at other points.

The feed material ordinarily charged into a plant contains a very small percentage of $U^{235}F_6$ with the remainder being $U^{238}F_6$. Consequently, a far greater amount of the feed flows downward in the cascade and is removed as tails than is removed at the top of the cascade as $U^{235}F_6$ or product. For this reason, as we progress upward in the cascade from the feed point, the total flow through the converters becomes progressively smaller. It follows, therefore, that in the ideal design each stage would be different in size from the adjacent stages. However, since the cost of construction of a production cascade is of great importance, a lower cost may be achieved by the use of many identical units. For this reason, the cascade is composed of a series of sections, each of which contains a large number of identical stages. The change in equipment size is accomplished in a step-wise manner by varying the equipment size of each so-called section.

Circuit Balances

Of major importance in a production cascade is the ability to calculate pressures, temperatures, and flows of the process gas which cannot be practically or economically measured by instrumentation. Since the cascade is made up of sections each of which contains a large number of identical stages and since the stage is the smallest separating unit of a cascade, these calculations are made on a stage basis. This system of calculating temperatures, pressures, and flows is commonly referred to as a circuit balance and consists of a series of calculations made from those pressures and temperatures ordinarily measured by instrumentation in the cascade, and from experimental data obtained from test loop studies.

For illustration purposes, the pressures, temperatures, and flows in a typical X-31 stage which are taken by instrumentation, plus those to be calculated, are shown in Figure A-3. It is obvious that the cost to install instruments in every stage to measure all of the values indicated would be prohibitive.

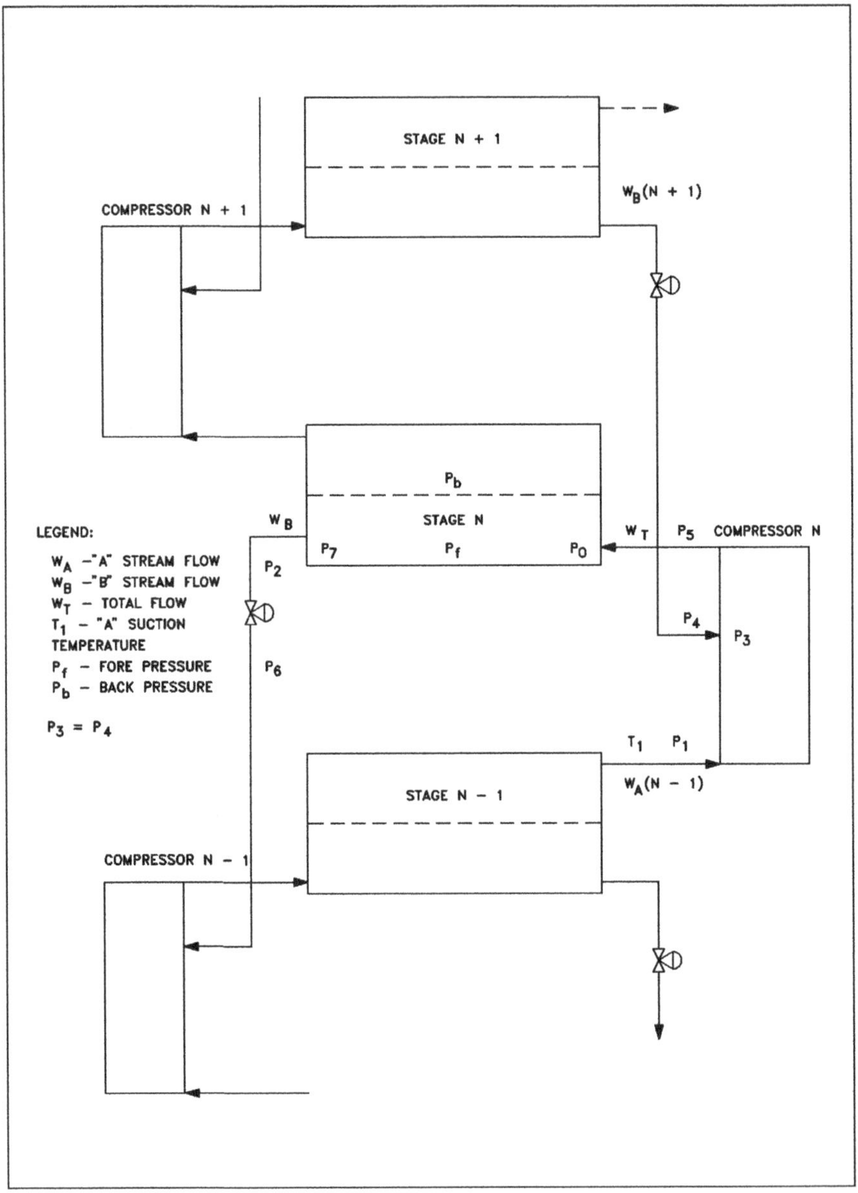

Figure A-3
Pressures, Temperatures, and Flows in a Typical V-31 Stage

Because of their number and complexity, the mathematical formulas used in calculating the pressure points in Figure A-3 will not be presented. The operator will read and record only those values which are measured by instruments and will not be required to perform any of the mathematical calculations involved in completing a circuit balance. However, the following discussion will give an indication as to the procedures involved in determining these values. To calculate a circuit balance for Stage N in Figure A-3, pressures P_1 and P_2 and temperature T_1 are recorded. Then, knowing P_1 and T_1 and a constant previously determined in a test loop, it is possible to calculate the flow W_A (N - 1). Since the "A" barrel of the compressor has essentially a constant suction volume in the normal operating range, the interstage flow is proportional to the compressor "A" suction pressure and temperature. Next, it is known that the flow in a cascade must operate at a steady state condition to prevent an accumulation of deficiency of inventory in any one section. It may therefore be assumed that the flow downward to a stage is equal to the flow upward from the same stage. (This assumption is not valid for all stages in a cascade but for practical applications the error is negligible.) From this assumption, W_B is equal to W_A (N - 1). With P_2, W_B, and pressure drop data obtained in the test loop, P_7 can be calculated.

In a test loop, data is taken on converters of different sizes and types to determine pressure variations with different amounts of flow and cut. If W_A is found in the same manner as W_A (N - 1), it is then possible with the test loop data to calculate P_b. Also, with the converter data, and knowing W_B, P_7, and W_T ($W_T = W_A + W_B$), P_o and subsequently, P_f are found.

Again using pressure drop data, P_5 can be calculated from W_T and P_o With P_5 P_3 is calculated by knowing the "B" pressure ratio of the compressor.

From the calculated values obtained from a circuit balance, it is possible to make further calculations which indicate the operational and productive characteristics of the cascade. For example, P_f, P_b, and the flows through a converter must be known before the actual permeability of the barrier can be calculated. Also, these values make it possible to accurately predict horsepower requirements under varying conditions of flow and pressure. Another important application of the circuit balance is in the determination of inventory in the cascade where pressures and temperatures not otherwise measured are required.

CONVERTERS

Converters

The actual separation of $U^{235}F_6$ from $U^{238}F_6$ in a gaseous diffusion plant is accomplished in the converter. The converter, sometimes referred to as a diffuser, contains the barrier tubing through which diffusion takes place. In the previous section we studied the flow through the converter as a part of a stage. In this section we will see what takes place inside the converter.

In all, the entire cascade contains over 4000 converters -- one in each stage. The stage size varies with the location in the cascade as shown in the table below.

TABLE A-1
Converter Stage Size vs. Location

Equipment Type	Converter Type	Length (inches)	Diameter (inches)	Number Used In Cascade
33	000	291	155	640
31	00-2	221	105	500
29	0	185 or 215	90 or 105	600
27	7	129	47	720
25	8	129	38	1560

Converter Construction

Externally, the converter resembles a large cylindrical tank resting on its side (see Figure A-4). There are openings at each end for the necessary piping connections. The outlet end of the converter contains the "A" and "B" stream process gas outlets. The other end contains the mixed process gas inlet and the stage coolant inlet and outlet. The cylindrical tank is called the shell and is constructed of steel with welded joints. Its internal surface is nickel plated. There are three external reinforcing flanges around the outside of the shell. The many hundreds of barrier tubes, or tube bundles, contained in one converter are held in place by the spool, or spool piece. The spool consists of struts and a central tube, or core, which is perforated and allows part of the "A" stream leaving the barrier tubes to reach the converter outlet. On either end of the core are tube sheets into the holes of which the barrier tubes are sealed by rolling or swagging. Tube sheets are also mounted on the core between the end tube sheets to support the barrier tubes.

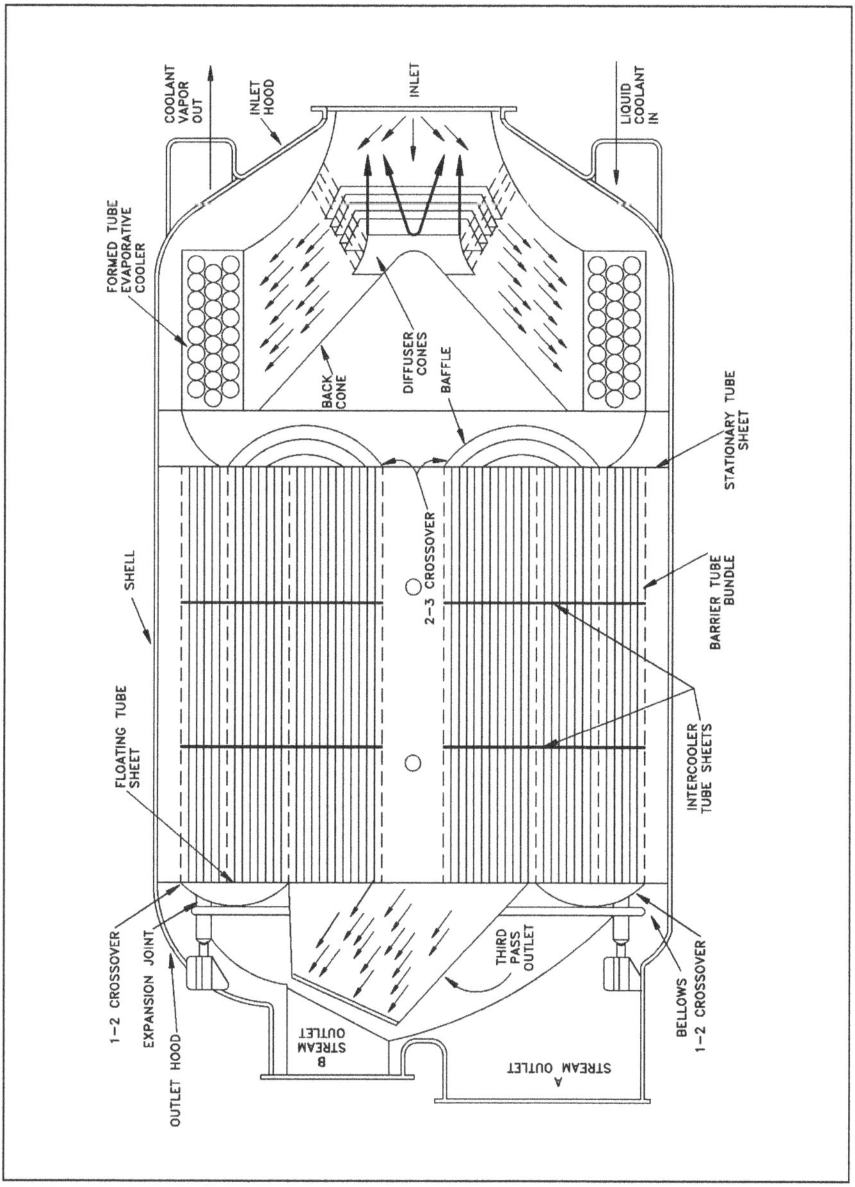

Figure A-4 - Typical Converter

The remainder of the converter equipment, with the exception of the gas cooler, is designed to direct the process gas flow inside the converter.

The Gas Cooler

The gas cooler's purpose is to remove the heat of compression from the process gas which has just been discharged from a compressor. Certain type of stages use a gas cooler contained within the converter as shown on Figure A-4. The gas coolers may be separate units placed between the compressors and the converters.

Barrier Tubing

In the early stages of converter design, it was recognized that barriers made in the form of flat sheets and stacked in a container would offer the most barrier area per unit volume. For practical reasons, the barrier had to be made in the form of tubes. More engineering knowledge was available for tubular construction and the urgency of the problem did not allow any unnecessary development time. The use of tubes also allowed more satisfactory replacement of damaged barrier material. As a result, the barrier is made in the form of thin-walled tubes.

Process Gas Flow

We can now trace the process gas flow within the converter. In the AB cooled converter of Figure A-4, the process gas enters one end of the converter at the center and is directed to the outside of the shell by a series of baffles. It flows through the gas cooler and is directed to the outer section, or pass, of the barrier tubes. All of the flow enters the outer, or first pass, and flows through the tubes. Part of the flow diffuses through the barrier tube walls and the remainder of the flow passes through the tubes and is directed by a crossover to the second pass of tubes. This crossover is sometimes called a doughnut because of its shape. The process gas flow which does not diffuse through the barrier tube walls of the second pass is directed by a second crossover to the third pass. The first crossover is sometimes refereed to as the 1-2 crossover because it directs gas flow from the first pass to the second pass. Similarly, the second crossover is called the 2-3 crossover. The undiffused process gas which passes through the third pass is directed to the "B" outlet of the converter and flows to the stage below. The process gas which was diffused through the barrier tubes in all three passes is collected and flows to the "A" outlet and on to the stage above.

Since some of the process gas flowing into the first pass will diffuse through the tube walls, a reduced amount of process gas flow will enter the second pass. In order to maintain the same flow velocity, the second pass contains fewer tubes than the first pass. Similarly, the third pass will contain still fewer tubes. The velocity of process gas through the tubes affects the flow through the tube walls and thus affects the separation efficiency. The efficiency is highest when the flow velocity is the same in all of the tubes.

In the Badger stage, about 50% of the process gas entering a converter diffuses through the tubing and flows to the stage above. This fraction is called the "cut." A cut of 50% has been found to yield the best separation. As mentioned previously, the "cut" in a Badger Cluster stage may be more or less than 50% depending upon its position in the cluster.

Diffusion

It is necessary to have a higher pressure inside the barrier tubes than outside in order to have a flow through the walls of the tubes. This inside pressure is called the fore pressure or high side pressure (H.S.P.), and is measured at an arbitrary point inside the tubes near the middle of the second pass. The pressure outside the tubes is the back pressure or low side pressure (L.S.P.), and is measured at an arbitrary point outside the tubes near the middle of the second pass.

The fore pressure is regulated by a control valve in the "B" stream, or down flow, from the converter. Actually, the control valve regulates the pressure immediately above it. This is called the control pressure and is the pressure which is indicated at the cell panel. It is slightly lower than the fore pressure due to the pressure drops in the converter and piping.

The rate of diffusion through the tubing walls for any given pressure drop across the barrier is determined by the permeability of the barrier. Mathematically, this is a dimensionless quantity which is the ratio of the rate of gas flow through the barrier to the rate of gas flow through the same area which would take place if the barrier were not there. The term, usually called permeability, is used to relate the actual permeability to the design permeability. For example, if the flow through the barrier has decreased 10% due to plugging of the barrier holes, the permeability would be 90%.

Intentionally Left Blank

Department of Energy
Fundamentals Handbook

CHEMISTRY
Module 2
Corrosion

TABLE OF CONTENTS

LIST OF FIGURES

LIST OF TABLES

REFERENCES

- Academic Program for Nuclear Plant Personnel, Volume II, Columbia, MD, General Physics Corporation, Library of Congress Card #A 326517, 1972.
- Dickerson, Gray, Darensbourg and Darensbourg, Chemical Principles, 4th Edition, The Benjamin Cummings Publishing Company, 1984.
- R. A. Day, Jr. and R. C. Johnson, General Chemistry, Prentice Hall, Inc., 1974.
- Glasstone and Sesonske, Nuclear Reactor Engineering, 3rd Edition, Van Nostrand Reinhold Company, 1981.
- Sienko and Plane, Chemical Principles and Properties, 2nd Edition, McGraw and Hill, 1974.
- Underwood, Chemistry for Colleges and Schools, 2nd Edition, Edward Arnold, Ltd., 1967.

TERMINAL OBJECTIVE

1.0 Without references, **DESCRIBE** the causes and effects of corrosion on metals and the type of chemistry used in a plant to minimize corrosion.

ENABLING OBJECTIVES

1.1 **DEFINE** the following terms:

a. Ionization
b. Conductivity
c. Corrosion
d. Electrolysis
e. General corrosion

1.2 **DESCRIBE** an electrochemical cell with respect to the corrosion of metals.

1.3 **STATE** what happens to a metal during the oxidation step of the oxidation-reduction process.

1.4 **STATE** what happens to a metal during the reduction step of the oxidation-reduction process.

1.5 **DEFINE** the following terms:

a. Passivity
b. Polarization

1.6 **DESCRIBE** the affects of passivity and polarization on the corrosion process.

1.7 **LIST** the two conditions that contribute to general corrosion.

1.8 **DESCRIBE** how the rate of corrosion occurring in the plant is affected by the following:

a. Temperature
b. Water velocity
c. Oxygen
d. pH
e. Condition and composition of the metal surface

ENABLING OBJECTIVES (Cont.)

1.9 **LIST** the three products that are formed from the general corrosion of iron.

1.10 **IDENTIFY** the action taken for initial fill of a reactor system to limit general corrosion.

1.11 **STATE** the four methods used to chemically control general plant corrosion.

1.12 **LIST** the six water chemistry conditions that limit corrosion of aluminum.

1.13 **DEFINE** the following terms:

a. Crud
b. Scale
c. Galvanic corrosion

1.14 **IDENTIFY** the five problems associated with the presence or release of crud into reactor coolant.

1.15 **STATE** the four causes of crud bursts.

1.16 **STATE** the two conditions that can cause galvanic corrosion.

1.17 **EXPLAIN** the mechanism for galvanic corrosion.

1.18 **IDENTIFY** the two locations that are susceptible to galvanic corrosion.

1.19 **STATE** the five control measures used to minimize galvanic corrosion.

1.20 **DEFINE** the following terms:

a. Pitting corrosion
b. Crevice corrosion
c. Stress corrosion cracking

1.21 **STATE** the two conditions necessary for pitting corrosion to occur.

1.22 **STATE** the particular hazard associated with pitting corrosion.

1.23 **STATE** the four controls used to minimize pitting corrosion.

ENABLING OBJECTIVES (Cont.)

1.24 **IDENTIFY** the three conditions necessary for stress corrosion cracking to occur.

1.25 **DEFINE** the term chemisorption.

1.26 **STATE** the hazard of stress corrosion cracking.

1.27 **STATE** the three controls used to prevent stress corrosion cracking.

1.28 **DESCRIBE** the two types of stress corrosion cracking that are of major concern to nuclear facilities including:

a. Conditions for occurrence
b. Method(s) used to minimize the probability of occurrence

Intentionally Left Blank.

CORROSION THEORY

Uncontrolled corrosion in DOE facilities can cause many serious problems. Corrosion occurs continuously, and every metal in every facility is subject to some type of corrosion. Even though corrosion cannot be eliminated, it can be controlled.

EO 1.1 DEFINE the following terms:

a. Ionization
b. Conductivity
c. Corrosion
d. Electrolysis
e. General corrosion

EO 1.2 DESCRIBE an electrochemical cell with respect to the corrosion of metals.

EO 1.3 STATE what happens to a metal during the oxidation step of the oxidation-reduction process.

EO 1.4 STATE what happens to a metal during the reduction step of the oxidation-reduction process.

EO 1.5 DEFINE the following terms:

a. Passivity
b. Polarization

EO 1.6 DESCRIBE the affects of passivity and polarization on the corrosion process.

Corrosion

In DOE facilities, especially reactor plants, many precautions are taken to control the corrosion of metals used in the various primary and secondary systems. Uncontrolled corrosion of the reactor systems is detrimental for the following reasons.

- Rapid localized corrosion may lead to penetration of the metal containing the coolant. Radioactive coolant would then leak from the system and jeopardize safe operation.

- Corrosion of the nuclear fuel cladding may cause the cladding to become brittle and less ductile. The swelling from the generation of fission gases within the fuel may then cause the cladding to crack or blister, and highly radioactive fission products may then be released to the coolant.

- Some of the metallic oxide corrosion products released to the coolant may be deposited on surfaces in the reactor core. The neutron flux in the core produces nuclear reactions in the corrosion products, and they become highly radioactive. Subsequently, these activated corrosion products may be released from the core and redeposited on surfaces outside the core region. The radiation fields resulting from this redeposited material may then significantly increase radiation levels, thereby complicating maintenance and access capabilities. The corrosion product oxides may also result in fouling of heat transfer surfaces and in the accelerated wear of moving parts by corrosion products trapped in or between them.

Current is the flow of electrons through a medium. An electric current can flow through a metal conductor, and the metal will not show any obvious chemical changes. This type of conduction of electricity is called metallic conduction.

Ionization is the process of adding electrons to or removing electrons from atoms or molecules, creating ions. High temperatures, electrical discharges, and nuclear radiation can cause ionization. Many metals have a tendency to lose electrons to atoms or ions that have a tendency to gain electrons. Current can be conducted by the movement of these ions. The compounds that conduct electric current by ion movement are called electrolytes, and this ionic motion is call electrolytic conduction. *Conductivity* is a measure of the ability of a substance to allow electron flow. In the context of corrosion, conductivity indicates the amount of ions in solution, which relates directly to the potential of corrosion taking place.

Corrosion is the deterioration of a material due to interaction with its environment. Corrosion can have many forms, both wet and dry. *Electrolysis* is the decomposition by electric current (in the context of corrosion the use of electrical current to bring about chemical change). This chapter will concentrate on the corrosion of metals in water-based environments (electrolytes). An electrolyte is defined as an electricity-conducting fluid; that is, it has positive and negative ions that can move and constitute an electrical current. Pure water has a limited number of dissociated H^+ and OH^- ions and is a relatively poor conductor of electricity. Addition of acids, bases, or salts that dissociate into ions increases the current-carrying capability of the water (electrolyte).

Electrochemical Cells

Corrosion is electrochemical in nature because the corrosive chemical reactions involve transfer of charge. Figure 1 shows the transfer of charge when iron is in contact with water or an acidic water solution.

Iron goes into solution as Fe^{++} ions. As these ions go into solution, the metal becomes negatively charged (by the electrons left behind) with respect to the electrolyte. A potential difference (voltage) is produced between the electrolyte and the metal. The process in which electrons are given up and positive metal ions are formed is called oxidation. The sites at which the oxidation takes place on the surface of the metal become electrochemical cells made up of micro-electrodes of the two different substances; the metal and the electrolyte.

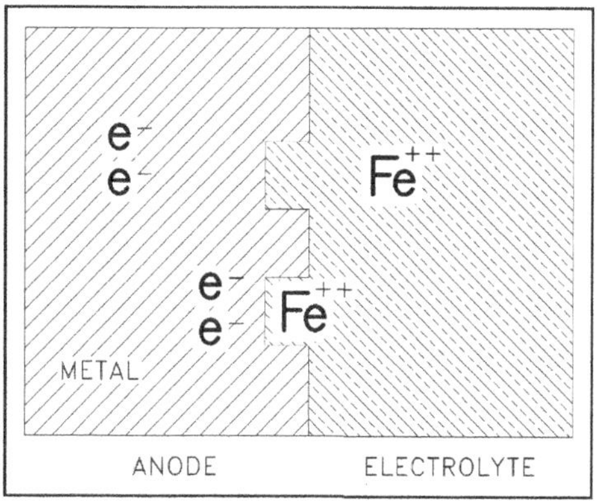

Figure 1 Formation of Ferrous (Fe^{++}) Ions in the Corrosion of Iron

These micro-electrodes set up many micro-cells connected through the bulk of the metal. If a different metal is used, it will go into solution to a greater (or lesser) extent producing a larger (or smaller) potential difference between the metal and electrolyte than was the case for iron. For example, magnesium and zinc go into solution to a greater extent than iron, and these metals will be more negative with respect to the electrolyte than iron. Nickel, lead, and copper go into solution less readily and produce a smaller potential difference. Table 1 lists the potential differences for various metals in water. The order of the series can change for different electrolytes (for example, different pH, ions in solution).

Electrochemical cells and oxidation potentials are very important in understanding most corrosion processes. Examples of electrochemical cells include galvanic cells (cells made up of electrodes of two different substances) and concentration cells (cells containing electrodes of the same substance under different conditions of concentration).

TABLE 1
Electromotive - Force Series (77°F)

Element	Electrode Reaction	Standard Electrode Potential, v
Sodium	$Na \rightarrow Na^{+} + e$	-2.712
Magnesium	$Mg \rightarrow Mg^{++} + 2e$	-2.34
Beryllium	$Be \rightarrow Be^{++} + 2e$	-1.70
Aluminum	$Al \rightarrow Al^{+++} + 3e$	-1.67
Manganese	$Mn \rightarrow Mn^{++} + 2e$	-1.05
Zinc	$Zn \rightarrow Zn^{++} + 2e$	-0.762
Chromium	$Cr \rightarrow Cr^{+++} + 3e$	-0.71
Iron	$Fe \rightarrow Fe^{+++} + 3e$	-0.44
Cadmium	$Cd \rightarrow Cd^{++} + 2e$	-0.402
Cobalt	$Co \rightarrow Co^{++} + 2e$	-0.277
Nickel	$Ni \rightarrow Ni^{++} + 2e$	-2.250
Tin	$Sn \rightarrow Sn^{++} + 2e$	-0.136
Lead	$Pb \rightarrow Pb^{++} + 2e$	-0.126
Hydrogen	$H \rightarrow 2H^{+} + 2e$	0.000 (reference)
Copper	$Cu \rightarrow Cu^{++} + 2e$	+0.345
Copper	$Cu \rightarrow Cu^{+} + e$	+0.522
Silver	$Ag \rightarrow Ag^{+} + e$	+0.800
Platinum	$Pt \rightarrow Pt^{++} + 2e$	+1.2
Gold	$Au \rightarrow Au^{+++} + 3e$	+1.42

The surface of any metal is a composite of a very large number of micro-electrodes, as illustrated in Figure 2. In order for corrosion to occur, the micro-cells must also be connected through some conducting path external to the metal. Usually the external connection is provided by water or an aqueous solution and the cells produce a current, allowing the chemical reactions responsible for corrosion to proceed.

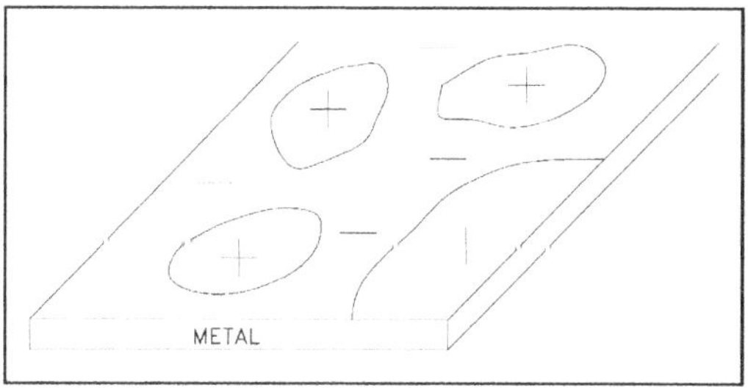

Figure 2 Metal Surface Showing Arrangement of Micro-cells

Consider iron in water again. If the surface of the iron and the water solution were uniform, iron would go into solution as Fe^{++} ions until the difference in potential between the positively-charged solution and the negatively-charged metal stopped the iron ions from leaving the surface. In practice, though, impurities and imperfections (for example, oxide coatings) lead to preferential removal of metal from certain parts of the surface, and potential differences arise as in the two metal system. The corrosion cells, changing as surface and solution differences change, cause general overall corrosion. If the cells do not shift, pitting results.

Oxidation-Reduction Reactions

The corrosion of a metal (that is, the chemical transformation that is recognized as destructive to the metal) is the oxidation step of the overall oxidation-reduction process. Oxidation is the process of losing electrons; reduction is the process of gaining electrons. The metal atoms release electrons (are oxidized) and become positive ions. The site at which this occurs is known as the anode. Typical oxidation half-reactions include the following.

$$Zn \rightarrow Zn^{+2} + 2e^- \quad (2\text{-}1)$$

$$Al \rightarrow Al^{+3} + 3e^- \quad (2\text{-}2)$$

$$Fe \rightarrow Fe^{+2} + 2e^- \quad (2\text{-}3)$$

The cations (positive ions) may then go into solution, or they may combine with any available anions (negative ions) or water to form ionic compounds. The exact fate of the cations is important to subsequent processes, but the primary effect is that atoms leave the metallic state, and the metal deteriorates.

An oxidation process cannot take place without a simultaneous reduction (gain of electrons) process. The nature of the reduction step in corrosion sometimes varies with the metal and the environment to which it is exposed. For most metals in an aqueous environment, the important reduction half-reaction is the reduction of hydronium ions (a hydronium ion is simply a hydrogen ion attached to a water molecule).

$$H_3O^+ + e^- \rightarrow H + H_2O \quad (2\text{-}4)$$

Small concentration variations within a solution in contact with the metal may also affect the rate and nature of corrosion reactions. Therefore, it is often impossible to predict the exact nature of corrosion reactions. It is generally found, however, that for most metals exposed to an aqueous environment the half-reactions involved in corrosion are the reduction reaction of Equation (2-4) and an oxidation half-reaction of the type shown in Equations (2-1) through (2-3).

General corrosion is the process whereby the surface of a metal undergoes a slow, relatively uniform; removal of material. This occurs on the surface of a single metal rather than dissimilar metals. In general corrosion, a nearly infinite number of micro-cells are established on the metal surface. Oxidation occurs at anodic areas and reduction at cathodic areas. The micro-cells are uniformly distributed over the metallic surface, and as the reaction proceeds the cells may migrate, or disappear and re-form. That is, any particular micro-region may be alternately anodic and cathodic. The result is a uniform attack on the metal surface.

Under some conditions, relatively large regions become anodic or cathodic. Such regions have less tendency to migrate and may remain operative for long periods of time. In this case, there will be severe attack of the metal at the anodic (oxidation) region. The result may be a visible pit in the metal surface.

Iron and steel are resistant to rapid corrosion in water despite the tendency of iron to oxidize as indicated by its standard electrode potential listed in Table 1. The reasons for this resistance are the passivating effect of the oxide film and cathodic polarization due to atomic hydrogen that absorbs on the oxide surface, both of which are explained in the next section.

Passivity and Polarization of Metal

Metals that normally fall victim to corrosion will sometimes exhibit a passivity to corrosion. *Passivity* is the characteristic of a metal exhibited when that metal does not become active in the corrosion reaction. Passivity is caused by the buildup of a stable, tenacious layer of metal oxide on the surface of the metal. This oxide layer is formed by corrosion on a clean metal surface, where the corrosion products are insoluble in the particular environment to which the metal is exposed. Once the layer, or film, is formed, it acts as a barrier separating the metal surface from the environment. For further corrosion to occur, the reactants must diffuse through the oxide film. Such diffusion is very slow or nonexistent, thus corrosion either decreases markedly or stops.

Metals such as zirconium, chromium, aluminum, and the stainless steels form thin, tenacious oxide films when exposed to the atmosphere or to pure water at room temperature. In some cases, the film is extremely thin and may be invisible to the unaided eye, but it is still very effective in giving these metals a marked passivity.

If there is a net conversion of reactants to products in a system, the system will be chemically unstable, and the reaction will continue until a stable state is attained. This stable state is known as equilibrium.

An active electrochemical cell (oxidation-reduction reaction) is an unstable chemical system. The potential associated with a galvanic cell, for example, steadily decreases as current flows and the oxidation-reduction reaction proceeds. Eventually, the potential falls to zero, the cell no longer supplies electrical energy, and no further net reaction takes place. At this point the system is at equilibrium. In electrochemical cells, the decrease in cell potential caused by the operation of the cell (current flow) is called *polarization.*

This change in cell potential can be determined. Consider the zinc-copper galvanic cell shown in Figure 3. As the reaction takes place, Zn^{+2} ions (produced by the oxidation of zinc metal) pass into solution. The Cu^{+2} ions in solution are reduced as the copper metal plates out. Thus, the concentration of Zn^{+2} in solution increases and the concentration of Cu^{+2} decreases according to the following overall reaction.

$$Zn + Cu^{+2} \rightarrow Zn^{+2} + Cu \qquad (2\text{-}5)$$

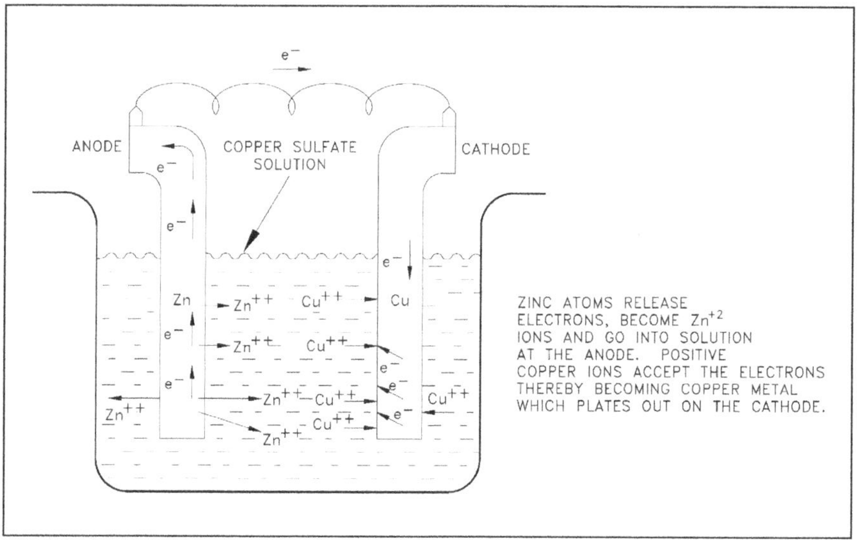

Figure 3 A Galvanic Cell

As Zn^{+2} increases and Cu^{+2} decreases, the electrical potential decreases. This decrease in cell potential, which results from changes in concentrations, is one form of polarization called concentration polarization.

Now consider a galvanic cell with zinc and platinum electrodes, such as that shown in Figure 4. The half-reactions in the cell are as follows.

$$Zn \rightarrow Zn^{+2} + 2e^-$$

$$H_3O^+ + e^- \rightarrow H + H_2O \quad (2\text{-}4)$$

Again, as the cell operates, the cell potential drops. The decrease is partially due to the increase in Zn^{+2} concentration and the decrease in H_3O^+ concentration, but another type of polarization also occurs in this cell. This second type is associated with the reduction half-reaction.

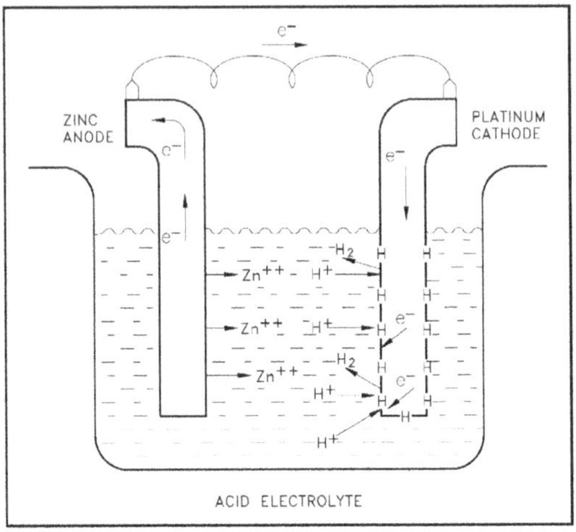

Figure 4 A Galvanic Cell Showing Absorbed Hydrogen Atoms on a Cathode

The hydrogen atoms formed by the reaction of Equation (2-4) absorb on the surface of the metal and remain there until removed by one of two processes: combination of two hydrogen atoms to form molecular hydrogen, which is then released as a gas or reaction with dissolved oxygen to form water. In the absence of oxygen (deaerated solutions), the first process applies.

$$2H \rightarrow H_2 \quad (2\text{-}6)$$

Combining Equation (2-6) with Equation (2-4), the net reduction half-reaction is obtained.

$$2(H_3O^+ + e^- \rightarrow H + H_2O)$$

$$2H \rightarrow H_2 \quad (2\text{-}6)$$

$$2H_3O^+ + 2e^- \rightarrow H_2 + 2H_2O \quad (2\text{-}7)$$

Until the absorbed hydrogen atoms are removed from the metal surface, they effectively block the sites at which the reaction of Equation (2-4) can occur. At low temperatures the reaction of Equation (2-6) is slow relative to the reaction of Equation (2-4) because, although the reaction is energetically favored, the combination of two hydrogen atoms requires a large activation energy. Equation (2-6) shows the rate-controlling step of the net reduction half-reaction. Because the oxidation half-reaction can occur no faster than the reduction half-reaction, the rate of the overall oxidation-reduction reaction is controlled by the reaction of Equation (2-6).

The layer of absorbed atomic hydrogen is said to polarize the cell. This type of polarization is called activation polarization and is sometimes referred to as hydrogen polarization, or cathodic polarization, because the polarizing reaction occurs at the cathode.

Both concentration and activation polarization decrease the net oxidation-reduction reaction rate. In corrosion processes, activation polarization usually has the greater effect.

Summary

The important information in this chapter is summarized below.

Corrosion Theory Summary

- Ionization is the process of adding electrons to or removing electrons from atoms or molecules which creates ions.
- Conductivity is a measure of the ability of a substance to allow electron flow.
- Corrosion is the deterioration of a material due to interaction with its environment.
- Electrolysis is the decomposition by electric current.
- General corrosion is the process whereby the surface of a metal undergoes a slow, relatively uniform, removal of material.
- Corrosion is electrochemical in nature because the corrosive chemical reactions involve a transfer of charge. The metal ions go into solution causing the metal to become negatively charged with respect to the electrolyte. The difference in the charge causes a potential to develop and produces a voltage between the electrolyte and the metal.
- The oxidation step of the oxidation-reduction process is where an atom (in this case a metal atom) releases electron(s) and becomes a positively-charged ion. The areas where oxidation takes place become electrochemical cells made up of two different substances. The oxidation step results in a chemical transformation that is destructive to the metal. The positive metal ions may go into solution, or they may combine with any available negative ions or water to form ionic compounds. An example of the oxidation step is:

$$Zn \rightarrow Zn^{+2} + 2e^-$$

Corrosion Theory Summary (Continued)

- The reduction step of the oxidation-reduction process is where a positively-charged ion gains electron(s). The reduction step for most metals, in an aqueous environment, is the reduction of hydronium ions. An example of the reduction step is:

$$H_3O^+ + e^- \rightarrow H + H_2O$$

- Passivity is the buildup of a stable, tenacious layer of metal oxide on the surface of the metal that acts as a barrier separating the metal surface from the environment. Passivity decreases or stops the corrosion process because of the formation of the layer.

- Polarization is the decrease in cell potential caused by the operation of the electrochemical cell. Polarization can be in two forms; concentration or activation. Concentration polarization is associated with the concentration of ions in solution which shields the metal, thereby causing a decrease in the electrical potential of the cell. Activation polarization is the formation of a layer containing absorbed hydrogen atoms that block the metal's surface from the corrosion process.

GENERAL CORROSION

This chapter describes the general corrosion processes of iron and carbon steel (not stainless steels) in aqueous environments. Of particular interest is the formation of the oxide film and the effects of system variables on the corrosion process.

EO 1.7 **LIST the two conditions that contribute to general corrosion.**

EO 1.8 **DESCRIBE how the rate of corrosion occurring in the plant is effected by the following:**

a. Temperature
b. Water velocity
c. Oxygen
d. pH
e. Condition and composition of the metal surface
f. Dissolved solids

EO 1.9 **LIST the three products that are formed from the general corrosion of iron.**

EO 1.10 **IDENTIFY the action taken for initial fill of a reactor system to limit general corrosion.**

EO 1.11 **STATE the four methods used to chemically control general plant corrosion.**

EO 1.12 **LIST the six water chemistry conditions that limit corrosion of aluminum.**

Conditions Contributing to General Corrosion

General corrosion is the process whereby the surface of a metal undergoes a slow, relatively uniform, removal of material. The two conditions typically required for a metal to undergo general corrosion are: 1) metal and water in the same environment, and 2) a chemical reaction between the metal and water that forms an oxide.

Corrosion of Iron

Unless noted otherwise, the following discussion applies to deaerated water at room temperature and approximately neutral pH. The affects of temperature, oxygen, and pH are discussed later in this chapter.

The oxidation and reduction half-reactions in the corrosion of iron are as follows.

$$Fe \rightarrow Fe^{+2} + 2e^- \quad \text{(oxidation)} \tag{2-3}$$

$$H_3O^+ + e^- \rightarrow H + H_2O \quad \text{(reduction)} \tag{2-4}$$

The overall reaction is the sum of these half-reactions.

$$Fe + 2H_3O^+ \rightarrow Fe^{+2} + 2H + 2H_2O \tag{2-8}$$

The Fe^{+2} ions readily combine with OH^- ions at the metal surface, first forming $Fe(OH)_2$, which decomposes to FeO.

$$Fe^{+2} + 2OH^- \rightarrow Fe(OH)_2 \rightarrow FeO + H_2O \tag{2-9}$$

Ferrous oxide (FeO) then forms a layer on the surface of the metal. Below about 1000°F, however, FeO is unstable and undergoes further oxidation.

$$2FeO + H_2O \rightarrow Fe_2O_3 + 2H \tag{2-10}$$

Atomic hydrogen then reacts to form molecular hydrogen, as described previously, and a layer of ferric oxide (Fe_2O_3) builds up on the FeO layer. Between these two layers is another layer that has the apparent composition Fe_3O_4. It is believed that Fe_3O_4 is a distinct crystalline state composed of O^{-2}, Fe^{+2}, and Fe^{+3} in proportions so that the apparent composition is Fe_3O_4. These three layers are illustrated in Figure 5.

Once the oxide film begins to form, the metal surface is no longer in direct contact with the aqueous environment. For further corrosion to occur, the reactants must diffuse through the oxide barrier. It is believed that the oxidation step, Equation (2-3), occurs at the metal-oxide interface. The Fe^{+2} ions and electrons then diffuse through the oxide layer toward the oxide-water interface. Eventually, Fe^{+2} ions encounter OH^- ions and form FeO. The electrons participate in the reduction reaction with hydronium ions. These latter reactions are believed to take place predominately at the oxide-water interface, but some reaction may occur within the oxide layer by the diffusion of H^+, OH^-, and H_2O into the layer.

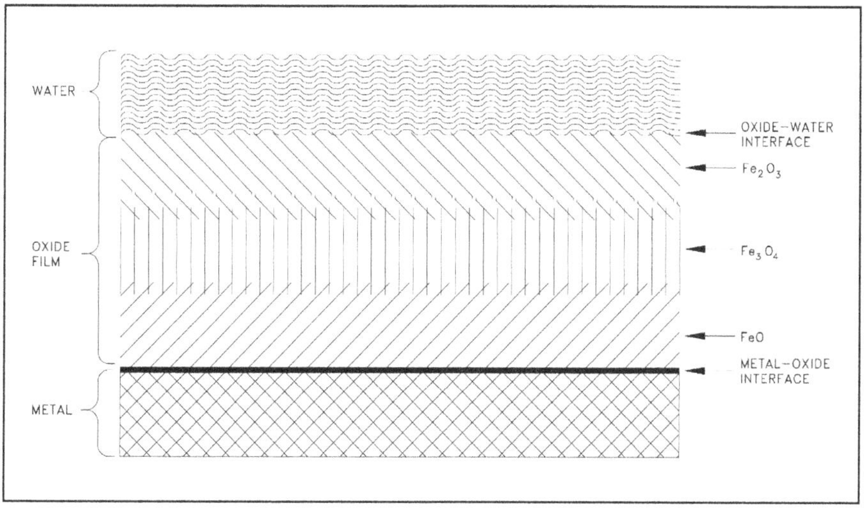

Figure 5 Simplified Schematic Diagram of Oxide Corrosion Film on the Surface of a Metal

Regardless of the exact diffusion mechanism, the oxide layer represents a barrier to continued corrosion and tends to slow the corrosion rate. The exact effect of this layer on the corrosion rate depends on the uniformity and tenacity of the film. If the film is loosely attached, develops defects, or is removed, the metal surface is again exposed to the environment and corrosion occurs more readily.

Factors Affecting General Corrosion Rate

Like most other chemical reactions, corrosion rates increase as temperature increases. Temperature and pressure of the medium govern the solubilities of the corrosive species in the fluid, such as oxygen (O_2), carbon dioxide (CO_2), chlorides, and hydroxides. A rule of thumb is that the reaction rate doubles with a 20°F to 50°F temperature rise. This linear increase with temperature does not continue indefinitely due, in part, to a change in the oxide film.

When water velocity is extremely high, the impact of the water tends to remove the protective oxide layer and some of the metal under it (erosion), thus, exposing more metal to corrosion. Water velocities of 30 to 40 ft per second are usually considered to cause erosion.

The presence of oxygen in water to which iron is exposed increases the corrosion rate. The reason for this increase is the rapid reaction between oxygen and the polarizing layer of atomic hydrogen absorbed on the oxide layer. The following reaction rapidly removes the polarizing layer.

$$O_2 + 4H \rightarrow 2H_2O \qquad (2\text{-}11)$$

The overall reaction can be obtained by combining Equations (2-4) and (2-11).

$$O_2 + 4H_3O^+ + 4e \rightleftharpoons 6H_2O$$

The controlling step is believed to be diffusion of O_2 to the metal surface where it can react directly with iron or with FeO.

$$2Fe + O_2 \rightarrow 2FeO \qquad (2\text{-}12)$$

$$4FeO + O_2 \rightarrow 2Fe_2O_3 \qquad (2\text{-}13)$$

Oxygen, therefore, has two effects: it removes the polarizing layer of atomic hydrogen, and it can react directly with the metal or metal oxide; thus, the corrosion rate increases. Substances, such as O_2 in this case, that remove the absorbed atomic hydrogen are called depolarizers. The depolarizing effect of O_2 is illustrated in Figure 6.

The effect of the pH of water to which iron or steel is exposed is influenced by temperature in the following manner. The potential of hydrogen or symbol (pH) is defined as the negative logarithm of the hydrogen concentration, represented as $[H^+]$ in moles/liter.

$$pH = -\log [H^+]$$

The pH value is used to represent the acidity of a solution.

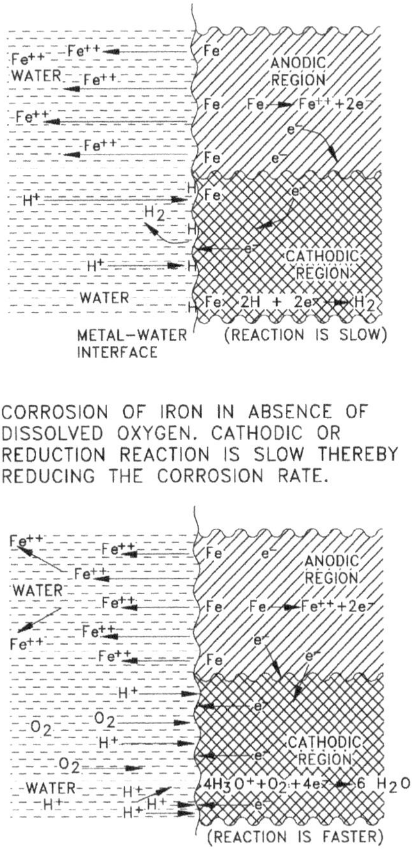

CORROSION OF IRON IN ABSENCE OF DISSOLVED OXYGEN. CATHODIC OR REDUCTION REACTION IS SLOW THEREBY REDUCING THE CORROSION RATE.

CORROSION OF IRON WITH DISSOLVED OXYGEN PRESENT IN WATER. RATE OF CATHODIC OR REDUCTION REACTION HAS INCREASED AS A RESULT OF OXYGEN DEPOLARIZATION. NET RESULT IS AN INCREASE IN CORROSION RATE.

Figure 6 Representation of Cathodic Depolarization by Oxygen

First, consider the exposure of iron to aerated water at room temperature (aerated water will contain dissolved oxygen). The corrosion rate for iron as a function of pH is illustrated in Figure 7. In the range of pH 4 to pH 10, the corrosion rate of iron is relatively independent of the pH of the solution. In this pH range, the corrosion rate is governed largely by the rate at which oxygen reacts with absorbed atomic hydrogen, thereby depolarizing the surface and allowing the reduction reaction to continue. For pH values below 4.0, ferrous oxide (FeO) is soluble. Thus, the oxide dissolves as it is formed rather than depositing on the metal surface to form a film. In the absence of the protective oxide film, the metal surface is in direct contact with the acid solution, and the corrosion reaction proceeds at a greater rate than it does at higher pH values. It is also observed that hydrogen is produced in acid solutions below a pH of 4, indicating that the corrosion rate no longer depends entirely on depolarization by oxygen, but on a combination of the two factors (hydrogen evolution and depolarization). For pH values above about pH 10, the corrosion rate is observed to fall as pH is increased. This is believed to be due to an increase in the rate of the reaction of oxygen with $Fe(OH)_2$ (hydrated FeO) in the oxide layer to form the more protective Fe_2O_3 (note that this effect is not observed in deaerated water at high temperatures).

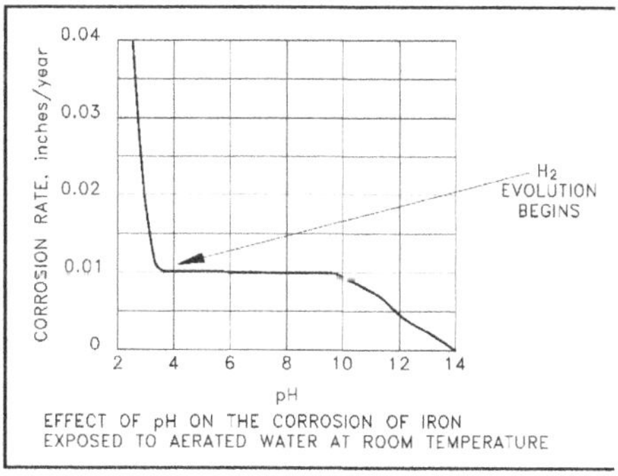

Figure 7 Effect of pH on the Corrosion Rate of Iron in Water

A plot of the relative corrosion rate for iron at various pH values in 590°F, oxygen-free water is presented as Figure 8. The curve illustrates that the corrosion rate of iron in high temperature water is lower in the pH range of 7 to 12 than it is at either lower or higher pH values (at very high pH values, greater than pH 13.0, the oxide film becomes increasingly more soluble because of increased formation of soluble FeO_2 at high temperatures, and corrosion rates increase). As a result of the data plotted in Figure 8 and other similar measurements, it is general practice to maintain high temperature water in the alkaline condition (but below very high pH values) to minimize the corrosion of iron and the steels exposed to the high temperature water.

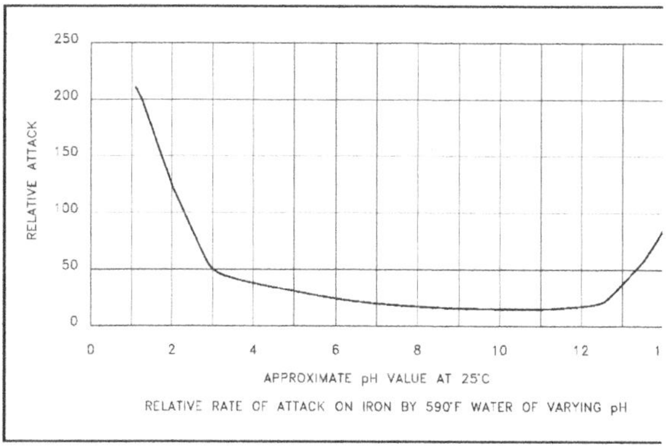

Figure 8 Effect of pH on the Relative Attack Rate of Iron in Water

The hydrogen normally dissolved in reactor coolant does not have any detectable direct effect upon the corrosion rate of the iron and steels exposed to reactor coolant. It does, however, have an important indirect effect by preventing the accumulation of dissolved oxygen in reactor coolant, which would accelerate corrosion. Dissolved oxygen reacts with the protective hydrogen gas layer at the cathode to form water.

The condition and composition of the metal surfaces affects the corrosion rate. Deposits, scale, or irregular surfaces create areas on the metal where local corrosion can initiate and proceed at a faster rate than normal. Certain alloys of metals have higher corrosion resistance than others, as discussed in the Material Science Handbook.

When iron or steel is exposed to high temperature water, the rate of corrosion of the metal is observed to decrease with exposure time during the early period of exposure. After a few thousand hours, the corrosion rate becomes relatively constant at a low value. During the early period of exposure, while the corrosion rate is decreasing, the oxide film on the surface of the metal grows in thickness. However, the rate at which the film grows decreases with time. The thickness of the oxide film soon reaches a relatively constant value, and thereafter film thickness does not change appreciably with further exposure. As might be expected, a relatively constant corrosion rate and oxide film thickness are attained at about the same time. Because a tightly adhering corrosion film inhibits further corrosion, great care is taken during the initial fill of reactor plants to promote formation of the best possible corrosion film. This process, referred to as pretreatment, or pickling, involves careful control of reactor coolant water chemistry and temperature during the pretreatment period.

Prevention Chemistry Control

Plant chemistry is used to control corrosion. The type of corrosion determines the method used for preventing or minimizing the corrosion rate.

Passivators and Inhibitors

Passivation is the condition where a naturally active metal corrodes at a very low rate, probably due to an oxide coating or an absorbed layer of oxygen. Some chemical substances, called passivators or inhibitors, if added to water, can provide this type of passivation by undergoing reduction at the metal surface. A common inhibitor is potassium chromate.

Cathodic Protection

The use of cathodic protection, supplying an external electric current to the iron so that it acts as a cathode and has no anodic areas, is another method of preventative chemical control. This can be accomplished by the use of an external voltage source or by the use of a sacrificial anode (e.g., zinc) which will corrode and provide the current.

Removing Corrosive Agents

Chemistry control in the form of removal of corrosive agents from a system is a widely used method. One method is using deaerators to remove dissolved oxygen and to a lesser extent carbon dioxide. Treating the water by softening and demineralization removes the dissolved solids and reduces the conductivity.

Chemical Addition

Chemical additions to a system that alter the chemical reaction or tie up a particular corrodant is a common method of control. Filming amines (organic compounds that are derivatives of ammonia) accomplish protection by forming adhering organic films on metal surfaces to prevent contact between corrosive species in the condensate and the metal surface. Phosphates and sodium hydroxide are used to adjust the system pH and remove hardness.

Corrosion of Aluminum

The corrosion of aluminum is dependent upon a vast number of variables. These variables include environment, temperature, alloy in question, flow velocities, impurities present in the environment, and chemistry conditions to which it is exposed. An additional factor that affects corrosion is pretreatment.

Many of the factors are controlled by design and construction, such as alloy type, temperature, and flow velocities. Pretreatment, soluble and solid impurities, and chemistry are within the control of the operator and will be discussed in this text.

Experiments have shown that prefilming limits corrosion on aluminum-clad fuel assemblies. In the tests conducted, the ratios of oxide film thickness for nonprefilmed and prefilmed elements were on the order of 2 to 3 and in some cases even greater.

Impurities are major contributors to the corrosion of aluminum. In most cases studied, the major source of contaminants has been makeup water systems. Corrosion products from other plant materials also contribute to the overall inventory of ionic and solid impurities. Finally, organic impurities from the resin used in ion exchangers have been detected. These occur in some installations because of the type of resin used and the particle filters normally installed on the ion exchanger effluents. This problem has been reduced by improved resins and installing filters capable of removing smaller particles.

Chemistry controls, including pH, dissolved oxygen, and conductivity, greatly influence the formation and propagation of the oxide film on aluminum surfaces. Dissolved oxygen is controlled for essentially the same reasons as for the corrosion of iron. Conductivity is a quantitative indication of the impurities present in the system, and pH theoretically dictates the value of conductivity.

For those reactor plants in which aluminum is used for cladding and other structural components, pH is controlled in an acidic condition because of the corrosion properties of aluminum. Plant pH has a marked effect on the rate of chemical reaction between the coolant water and aluminum. In the area of the cladding, the corrosion reduces the thickness and forms an oxide film that is a thermal barrier. Extensive tests carried out in support of DOE test reactors have revealed that minimum aluminum corrosion results with a pH of 5.0 at normal operating temperatures. Additionally, studies have shown that the aluminum corrosion products also exhibit a minimum solubility at a pH near 5.5 at 25 °C. The aluminum corrosion products tend to reduce the substrate (base) aluminum metal corrosion rates. Because it is desirable to maintain dissolved aluminum in the reactor coolant at the lowest practicable level, it is desirable to maintain the system pH level in the range of minimum oxide solubility. Figure 9 shows the effect of pH on aluminum oxide solubilities for various forms of oxide, and the effect of pH on corrosion rates. It should be noted that the values at which minimum corrosion and solubility are found shift to a lower pH as the temperature is increased. For example, at 300 °C, the value for minimum aluminum corrosion is near pH 3.0. Therefore, the optimum pH for operation is determined by the operating temperature.

The conditions that have proven to be most effective in limiting corrosion of aluminum are as follows.

- Maintaining pH slightly acidic with the value of the pH depending largely upon operating temperature
- Elimination of dissolved oxygen
- Elimination of soluble and solid impurities
- Prevention of the introduction of organic impurities
- Pretreatment (or pickling)
- Maintaining water purity

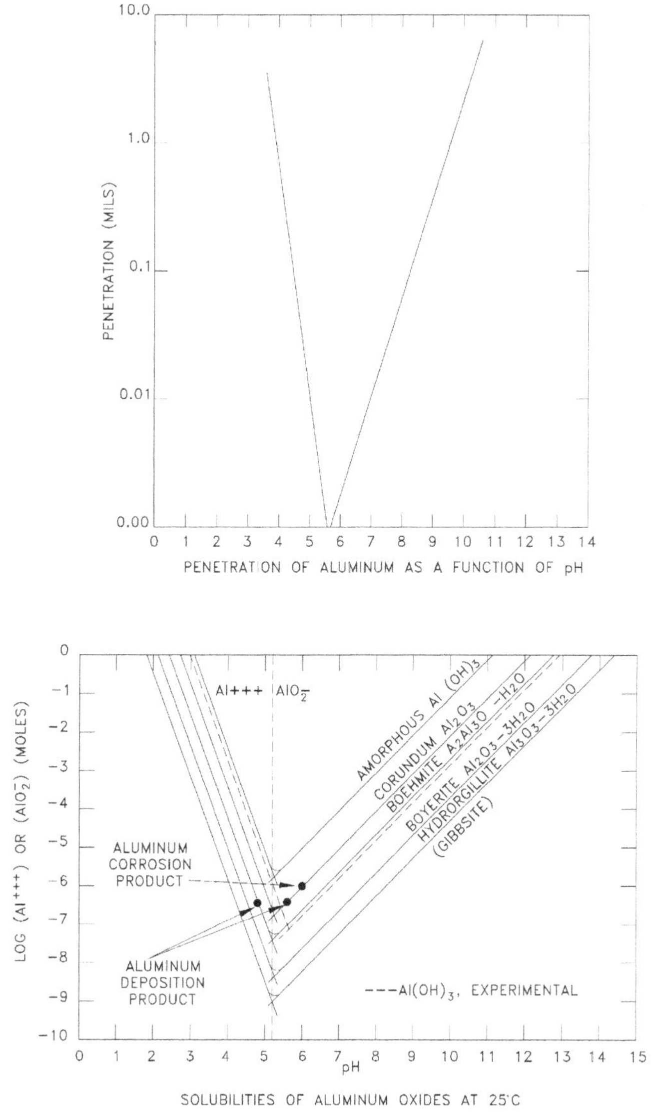

Figure 9 Effect of pH on Aluminum Corrosion and Oxide Solubility

Summary

The major points of this chapter are summarized below.

General Corrosion Summary

- The two conditions that contribute to general corrosion are:

  Metal and water in the same environment
  Chemical reaction between the metal and water to form an oxide

- The corrosion rate is affected by the following:

  A temperature rise in the range of 20°F to 50°F doubles the corrosion rate until the formation of the protective oxide film is complete.

  An extremely high water velocity, 30 to 40 ft per second, tends to remove the oxide film allowing the corrosion rate to increase.

  The presence of oxygen in water to which iron is exposed increases the corrosion rate. The reason for the corrosion rate increase is due to the rapid reaction between the oxygen and the polarizing layer of hydrogen absorbed on the oxide layer.

  A pH between 4 and 10 results in minimal corrosion rate. If the pH falls below or above this range, the corrosion will normally increase.

  The condition and composition of the metal surfaces affects the corrosion rate. Deposits, scale, or irregular surfaces create areas on the metal where local corrosion can initiate and proceed at a faster rate than normal.

  Dissolved solids tend to make it easier for current to flow, which results in a higher corrosion rate.

- The three products formed from general corrosion of iron are FeO, Fe_2O_3, and Fe_3O_4.

General Corrosion Summary (Cont.)

- During the initial fill of a reactor system, great care is taken to promote the formation of the best possible protective corrosion film. The protective film acts as a barrier and stops further uncontrolled corrosion from taking place. This process, called pretreatment or pickling, involves careful control of reactor coolant water chemistry and temperature during the pretreatment period.

- Four methods used to chemically control general plant corrosion are:

  Passivators (inhibitors) consist of a naturally active metal that corrodes at a very low rate. A common passivator is potassium chromate.

  Cathodic protection is the external application of an external electric current to the iron so that it acts as a cathode and has no anodic areas. This is accomplished by using either an external electrical source or the use of a sacrificial anode to provide the electrical current.

  Removing corrodants to reduce the dissolved oxygen and lesser extent carbon dioxide. Treating the water in this manner reduces the dissolved solids and reduces the conductivity.

  Chemical addition to alter the chemical reaction or tie up a particular corrodant. Filming amines, phosphates, and sodium hydroxide are used to provide organic films and adjust the system pH.

- The six water chemistry conditions that limit corrosion of aluminum are:

  Maintaining pH slightly acidic with the value of the pH depending largely on operating temperature.

  Elimination of dissolved oxygen

  Elimination of soluble and solid impurities

  Prevention of the introduction of organic impurities

  Pretreatment (or pickling)

  Maintaining water purity

CRUD AND GALVANIC CORROSION

A major potential problem is crud. Crud can cause an increase in background radiation levels. When two dissimilar metals are in close vicinity, the opportunity for ion transfer is high. The result is a perfect environment for galvanic corrosion. Because of the numerous types of materials used in nuclear facilities, galvanic corrosion is a major concern.

EO 1.13 DEFINE the following terms:

- **a. Crud**
- **b. Scale**
- **c. Galvanic corrosion**

EO 1.14 IDENTIFY the five problems associated with the presence or release of crud into reactor coolant.

EO 1.15 STATE the four causes of crud bursts.

EO 1.16 STATE the two conditions that can cause galvanic corrosion.

EO 1.17 EXPLAIN the mechanism for galvanic corrosion.

EO 1.18 IDENTIFY the two locations that are susceptible to galvanic corrosion.

EO 1.19 STATE the five control measures used to minimize galvanic corrosion.

Crud

In addition to the corrosion film, corrosion products in the form of finely divided, insoluble oxide particles called *crud* become suspended in the reactor coolant or loosely adhere to metal surfaces. Crud has several undesirable characteristics. It can be transported throughout the reactor coolant system. As a result, it can accumulate and foul heat-transfer surfaces or clog flow passages. The most undesirable characteristic of crud, however, is that it becomes activated when exposed to radiation. Because crud can be transported throughout the reactor coolant system, it can collect outside the reactor core, causing radiation hot spots that increase ambient radiation levels. Hot spots caused by collections of crud may occur at the entrance to the purification heat exchanger and other areas of low flow velocity. Crud that is loosely adhered to metal surfaces can suddenly become suspended in the reactor coolant.

The crud release can result from an increased oxygen concentration, a reduced (or significantly changed) pH, a large temperature change (heatup or cooldown), or a physical shock to the system. Physical shocks include starting, stopping, or changing pump speeds, or other evolutions like a reactor scram or a relief valve lift. The result is a sudden increase in reactor coolant activity. The release of crud in this fashion is termed a crud burst. Crud bursts often lead to the removal of protective corrosion films and make the freshly exposed metal more susceptible to additional corrosion. In addition to the corrosion film and crud, some of the corrosion products are soluble and are easily transported throughout the system.

High crud concentrations in the system can also complicate disposal of primary coolant. Many of the corrosion products have relatively long half-lives and represent significant biological hazards. If, therefore, primary coolant is drained or leaks from the plant shortly after a crud burst, additional procedures may need to be utilized to minimize the effects of this condition.

Therefore, if the conditions mentioned previously (O_2, pH) are changed, the solubility of these corrosion products will change, and they can then be transported to and deposited anywhere in the reactor coolant system.

Another corrosion byproduct is *scale*, which is made up of deposits on surfaces from the formation of insoluble compounds from normally soluble salts. Most common are calcium or magnesium carbonates ($CaCO_3$ or $MgCO_3$).

Galvanic Corrosion

Galvanic corrosion is the corrosion that results when two dissimilar metals with different potentials are placed in electrical contact in an electrolyte.

Of all the different types of corrosion, galvanic corrosion corresponds most closely to the electrochemical cells described previously in this module because galvanic corrosion occurs when two electrochemically dissimilar metals are joined together (in electrical contact) in a conducting medium (electrolyte). It may also take place with one metal with heterogeneities (dissimilarities) (for example, impurity inclusions, grains of different sizes, difference in composition of grains, differences in mechanical stress); abnormal levels of pH; and high temperatures. A difference in electrical potential exists between the different metals and serves as the driving force for electrical current flow through the corrodant or electrolyte. This current results in corrosion of one of the metals. The larger the potential difference, the greater the probability of galvanic corrosion. Galvanic corrosion only causes deterioration of one of the metals. The less resistant, active metal becomes the anodic corrosion site. The stronger, more noble metal is cathodic and protected. If there were no electrical contact, the two metals would be uniformly attacked by the corrosive medium as if the other metal were absent. Two locations susceptible to galvanic corrosion is a piping transition from one metal to another and a sacrificial anode (such as zinc).

Figure 10 illustrates that galvanic corrosion occurs when two different metals are in contact and exposed to an electrolyte.

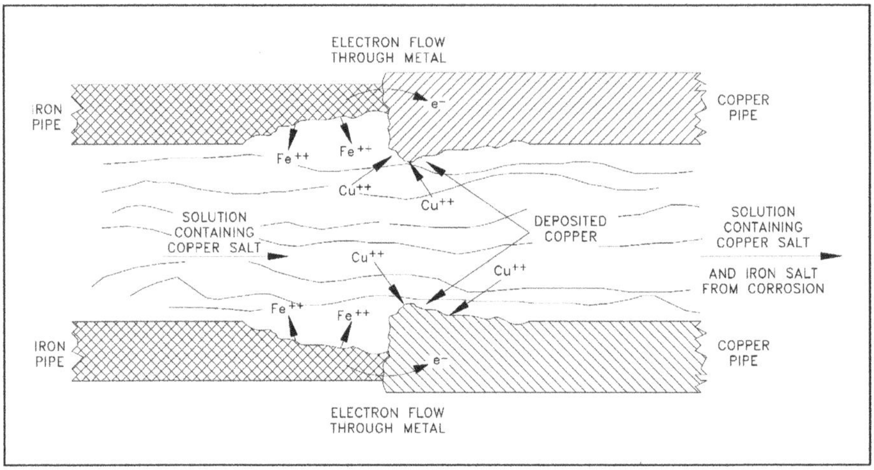

Figure 10 Galvanic Corrosion at Iron-Copper Pipe Junction

Figure 10 shows the junction of iron and copper pipes containing a solution of a copper salt. The oxidation potential of iron is sufficiently greater than that of copper so that iron is capable of reducing Cu^{+2} ions to copper metal. In this case, iron corrodes near the junction, and additional copper builds up on the copper pipe near the junction.

The solution to which the metal junction is exposed need not contain a salt of one of the metals for galvanic corrosion to occur. If the iron-copper junction were exposed to water without Cu^{+2} ions, the reduction reaction would be as shown in Equation (2-4).

$$H_3O^+ + e^- \rightarrow H + H_2O \tag{2-4}$$

Again, iron would corrode near the junction, but in this case hydrogen would be formed on the surface of the copper.

Prevention of Galvanic Corrosion

A method called cathodic protection, discussed previously in this module, is often used to retard or eliminate galvanic corrosion. One of several ways of accomplishing this is to attach a third metal to the metals to be protected. This metal must have an oxidation potential even greater than that of the metal to be protected. The most active metal then tends to corrode in place of the protected metal. The metal that corrodes to protect another metal is called a sacrificial anode. This method is applied in the original design of structural materials. Zinc is a common sacrificial anode and is often used in cooling water systems that contain seawater.

Galvanic corrosion can also be limited by: 1) using only metals that are close on the activity series (discussed in the chapter on *Corrosion Theory*), 2) electrical insulation of dissimilar metals, and 3) using poorly-conducting electrolytes (very pure water).

The relative surface areas of the two metals are also important. A much larger surface area of the non-active metal, compared to the active metal, will accelerate the attack. It has been determined that the relative surface area is the determining factor in the corrosion rates.
The required electrical current for galvanic corrosion will be stopped if the dissimilar metals are:

- separated by a non-conducting junction,
- separated from a conductive environment, and
- located in a poorly conducting electrolyte (pure water).

Summary

The important information of this chapter is summarized below.

Crud and Galvanic Corrosion Summary

- Crud is corrosion products in the form of finely divided, insoluble oxide particles suspended in the reactor coolant or loosely adhered to metal surfaces or activated corrosion and wear products.

- Scale is the deposition on the surfaces of the piping from the formation of insoluble compounds from normally soluble salts. Most common are calcium or magnesium carbonates.

- Galvanic corrosion is the corrosion that results when two dissimilar metals with different potentials are placed in electrical contact in an electrolyte.

- The problems of crud in reactor plants are:

  Fouling of coolant flow paths
  Fouling of heat transfer surfaces
  High general background (ambient) radiation levels
  Radiation hot spots
  Radioactive waste disposal

Crud and Galvanic Corrosion Summary (Cont.)

- The causes of a crud burst in the reactor coolant are:

  Increased oxygen concentration
  Reduced (or significantly changed) pH
  Large temperature change
  Physical shock (for example, starting and stopping pumps, changing speeds of pumps, reactor scram, or relief valve lift)

- Galvanic corrosion functions on the principle of the electrochemical cell, and occurs when two electrochemically dissimilar metals are joined together in a conducting medium. The two dissimilar metals generate an electrical potential, and this electrical potential serves as the driving force for the electrical current flow through the corrodant or electrolyte. The less resistant metal, called the active metal, becomes anodic. The other metal, called the noble metal, becomes cathodic.

- The two locations susceptible to galvanic corrosion are piping transitions between two dissimilar metals and at sacrificial anodes.

- Measures used to control galvanic corrosion include:

  Cathodic protection by introducing a third metal (sacrificial anode, normally zinc) to the metals being protected or using only metals that are close on the activity series.

  Choosing relative surface areas such that the material to be protected has a larger surface area than the active metal.

  Separating dissimilar metals with a non-conducting material

  Separating the metals from a conductive environment

  Use of poorly conducting electrolytes (pure water)

SPECIALIZED CORROSION

The environment in which a metal exists has direct bearing on the corrosion rate of that metal. Because of the unique environment possible in the nuclear industry, there are a few specialized types of corrosion that must be considered.

EO 1.20 DEFINE the following terms:

- **a. Pitting corrosion**
- **b. Crevice corrosion**
- **c. Stress corrosion cracking**

EO 1.21 STATE the two conditions necessary for pitting corrosion to occur.

EO 1.22 STATE the particular hazard associated with pitting corrosion.

EO 1.23 STATE the four controls used to minimize pitting corrosion.

EO 1.24 IDENTIFY the three conditions necessary for stress corrosion cracking to occur.

EO 1.25 DEFINE the term chemisorption.

EO 1.26 STATE the hazard of stress corrosion cracking.

EO 1.27 STATE the three controls used to prevent stress corrosion cracking.

EO 1.28 DESCRIBE the two types of stress corrosion cracking that are of major concern to nuclear facilities including:

- **a. Conditions for occurrence**
- **b. Method(s) used to minimize the probability of occurrence**

Pitting and Crevice Corrosion

Another possible effect of dissolved oxygen is accelerated localized attack. This is especially likely in areas of limited circulation. The resulting corrosion is called pitting corrosion.

Pitting corrosion occurs where the anodic site becomes fixed in a small area and the formation of holes (deep attack) in an otherwise unaffected area takes place. *Crevice corrosion* is a type of pitting corrosion that occurs specifically within the low flow region of a crevice.

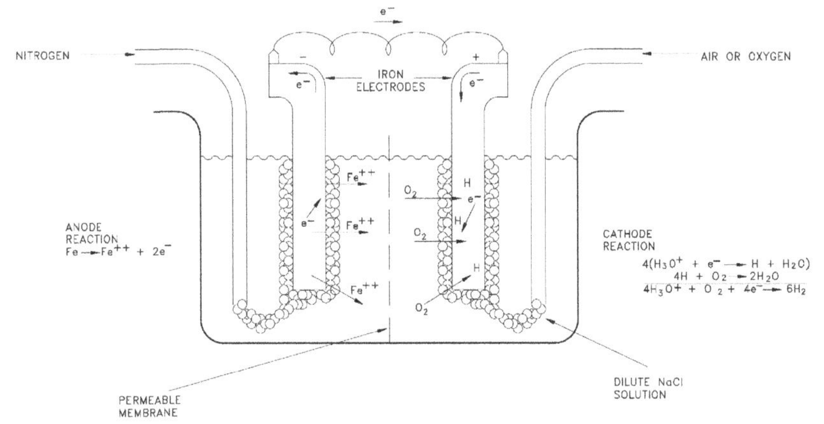

Figure 11 Differential Aeration Cell

To illustrate pitting attack, consider a special type of galvanic cell called a differential aeration cell such as the one illustrated in Figure 11. This particular differential aeration cell is showing current flow as a result of depolarization of one electrode (cathode) by oxygen. In this type of cell, two iron electrodes are exposed to a dilute solution of an electrolyte (NaCl, for example). Air (or oxygen) is bubbled around one electrode, and nitrogen is bubbled around the other. A current flows through the wire connecting the two electrodes. The difference in potential is a result of the difference in oxygen concentration at the two electrode surfaces. At the electrode exposed to nitrogen, electrons are given up by the iron as it is oxidized. These electrons readily flow through the external circuit to the electrode exposed to oxygen. At this depolarized electrode they can participate in a reduction reaction. As a result, oxidation occurs at the electrode exposed to nitrogen and reduction occurs at the aerated electrode. Oxidation at one electrode and reduction at the other creates a potential and a flow of current through the connecting wire. Note that loss of metal occurs at the electrode that is deficient in oxygen.

In iron that is exposed to water, a similar action can occur if adjacent areas of the metal surface become exposed to solutions with different oxygen concentrations. For example, the solution in a crevice exchanges slowly with the bulk of the solution outside the crevice. Oxygen in the solution inside the crevice will be depleted initially by the corrosion reaction.

$$2Fe + O_2 \rightarrow 2FeO \tag{2-12}$$

This reaction alone does not produce a protective film on the metal. Because of restricted flow into the crevice, replenishment of oxygen will be very slow; therefore, the solution inside the crevice will have a low oxygen concentration relative to that outside the crevice as shown in Figure 12. The two adjacent areas then establish a concentration cell with electrons flowing from the region of low oxygen concentration to the region of high concentration. Thus, metal goes into solution (oxidation) inside the crevice, and reduction occurs outside the crevice. Metal ions diffuse out of the crevice, more metal dissolves, and the process continues. This results in the formation of a pit inside the crevice.

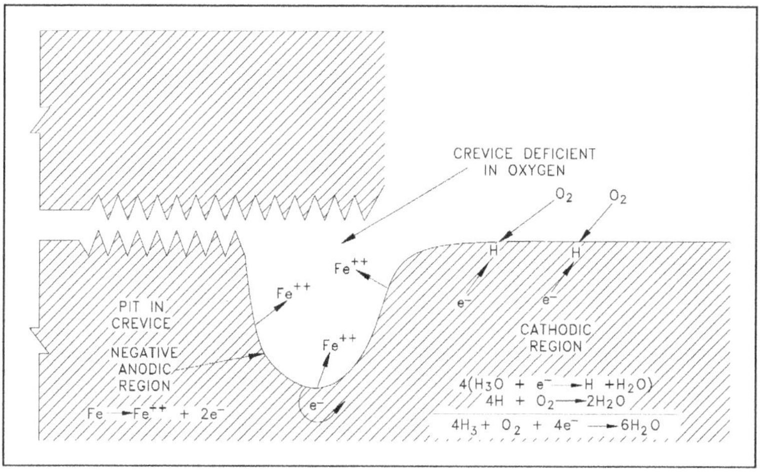

Figure 12 Representation of Crevice Pitting

The presence of oxygen can also promote pitting at areas on the metal surface that are initially anodic with respect to an adjacent area. For example, suppose that adjacent areas on a metal surface exhibit slightly different oxidation potentials. Oxidation, or loss of metal, proceeds at the region of higher potential. Corrosion in the region of higher potential leads to formation (at least initially) of a porous oxide film. The thickness of the film formed on the adjacent cathodic region will be much less. Oxygen in the bulk of solution can reach the cathodic surface (with the thin film) more readily than it can the nearby anodic surface region (with the thicker oxide film). Depolarization of the cathodic region (thin film) by oxygen tends to maintain this region cathodic, while a deficiency of oxygen under the thicker porous corrosion film assists in maintaining an anodic condition in this region. The overall result is corrosion, or wasting away, of the metal in the anodic region under the thicker film. Thus, a pit in the metal surface is formed under the mound of surface oxide, as illustrated in Figure 13. Pitting of this type is common in both low temperature and high temperature iron-water systems if precautions are not taken to remove the oxygen from the water within the system.

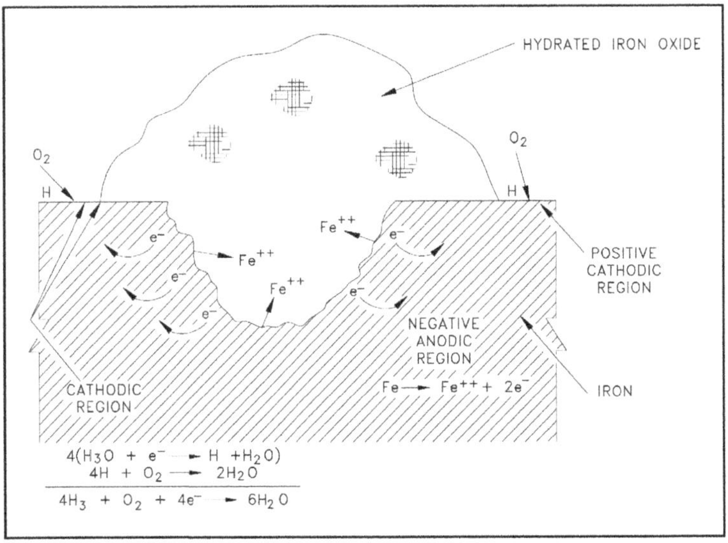

Figure 13 Pit in Metal Surface Promoted by Depolarization

It is also found that certain ions, notably chloride ions, cause pitting of iron and steel. The exact mechanism by which this occurs is not clear, but in some way chloride ions cause defects in the passivating oxide layer on the metal surface. The defects are highly localized and are surrounded by large passive areas that tend to be cathodic. Thus, a small anodic (oxidation) site is surrounded by a large cathodic (reduction) area. The current density will then be very large at the anodic site, and attack on the metal will be rapid. In some test cases, deep pits have been observed within a few hours.

Pitting and crevice corrosion are a major hazard to a nuclear facility because of the rapid penetration of the metal with little overall loss of mass. A nuclear facility minimizes pitting and crevice corrosion by the following actions.

- Avoiding stagnant or low flow conditions.
- Using metals and alloys that are less susceptible to the corrosion.
- Avoiding agents in the medium that cause pitting (for example, chlorides and oxygen).
- Designing the system and components such that no crevices are present.

Stress Corrosion Cracking

Stress corrosion cracking (SCC) is a type of intergranular attack corrosion that occurs at the grain boundaries under tensile stress. Grain boundaries are discussed in detail in the *Material Science* Handbook. SCC occurs in susceptible alloys when the alloy is exposed to a particular, specific environment if the alloy is in a stressed condition. Stress corrosion cracking appears to be relatively independent of general uniform corrosion processes. Thus, the extent of general corrosion can be essentially nil, and stress cracking can still occur. Most pure metals are immune to this type of attack.

According to the most widely accepted theory, stress corrosion cracking is caused by a process called chemisorption. Unlike relatively weak physical absorption, such as hydrogen gas on platinum metal, chemisorption may be thought of as the formation of a compound between the metal atoms on the surface as a monomolecular layer of the chemisorbed substance, such as Cl-, OH-, Br-, and some other ions. The formation of this chemisorbed layer greatly reduces the attraction between neighboring metal atoms. A defect initially present then grows as the metal atoms separate under stress, more chemisorption occurs, and the process continues. In very severe cases, the time required for this cracking to occur is only a matter of minutes.

Many stainless steels are susceptible to stress corrosion cracking. Stainless steels containing 18 percent chromium and 8 percent nickel are susceptible to cracking in environments containing chloride ions and in concentrated caustic environments (that is, in environments where the hydroxyl ion concentration is high). On the other hand, these types of stainless steels do not exhibit any tendency to crack when they are exposed to water environments containing nitrate (NO_3^-), sulfite (SO_3^{--}), and ammonium (NH_3^+) ions.

SCC is of great concern because it can readily crack metal of appreciable thickness. If the environment is severe enough, cracking can occur in a very short period of time. The crack can then lead to a serious failure of the component, or the system, and all the attendant results (for example, contamination, loss of coolant, and loss of pressure).

The most effective means for preventing SCC are proper design, reducing stress, removing critical environmental contributors (for example, hydroxides, chlorides, and oxygen), and avoiding stagnant areas and crevices in heat exchangers where chlorides and hydroxides might become concentrated. Low alloy steels are less susceptible than high alloy steels, but they are subject to SCC in water containing chloride ions. Nickel based alloys are not affected by chloride or hydroxide ions.

Two types of SCC are of major concern to a nuclear facility.

Chloride Stress Corrosion Cracking (Stainless Steels)

The three conditions that must be present for chloride stress corrosion to occur are as follows.

- Chloride ions are present in the environment
- Dissolved oxygen is present in the environment
- Metal is under tensile stress

Austenitic stainless steel is a non-magnetic alloy consisting of iron, chromium, and nickel, with a low carbon content. This alloy is highly corrosion resistant and has desirable mechanical properties. One type of corrosion which can attack austenitic stainless steel is chloride stress corrosion. Chloride stress corrosion is a type of intergranular corrosion.

Chloride stress corrosion involves selective attack of the metal along grain boundaries. In the formation of the steel, a chromium-rich carbide precipitates at the grain boundaries leaving these areas low in protective chromium, and thereby, susceptible to attack. It has been found that this is closely associated with certain heat treatments resulting from welding. This can be minimized considerably by proper annealing processes.

This form of corrosion is controlled by maintaining low chloride ion and oxygen content in the environment and the use of low carbon steels. Environments containing dissolved oxygen and chloride ions can readily be created in auxiliary water systems. Chloride ions can enter these systems via leaks in condensers or at other locations where auxiliary systems associated with the nuclear facility are cooled by unpurified cooling water. Dissolved oxygen can readily enter these systems with feed and makeup water. Thus, chloride stress corrosion cracking is of concern, and controls must be used to prevent its occurrence.

Figure 14 illustrates intergranular stress corrosion cracking. The pressure of a tensile stress opens up intergranular cracks and accelerates further corrosion. Chloride stress corrosion is a particularly significant problem in the operation of nuclear facilities because of the wide use of austenitic stainless steel, and the inherent presence of high tensile stresses associated with pressurization. Chloride stress corrosion cracks have been known to propagate in austenitic stainless steel at stresses of about one-fifth yield strength with chloride concentrations of less than 50 ppm. Yield strength is discussed in detail in the *Material Science* Handbook.

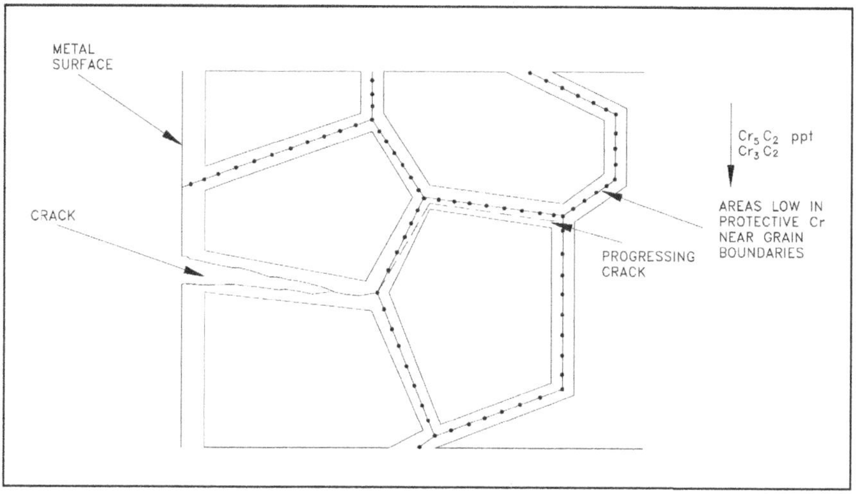

Figure 14 Intergranular Corrosion Cracking

Tests show that the 18-8 stainless steels are susceptible to chloride stress attack when both the chloride ion concentration and dissolved oxygen concentration are above certain values. The region of susceptibility for austenitic stainless steel is illustrated in Figure 15. Note that when dissolved oxygen is present at about 1 ppm, chloride stress corrosion cracking can be initiated at chloride ion concentrations near 1 ppm. However, when the concentration of dissolved oxygen is very low, susceptibility to chloride stress corrosion cracking is reduced.

High temperature tends to decrease the time required for chloride-induced cracking to occur, but there appears to be no practical temperature limit below which cracking will not occur, given sufficient time and severe conditions. The curve in Figure 15 is valid for temperatures in the range 470°F to 500°F.

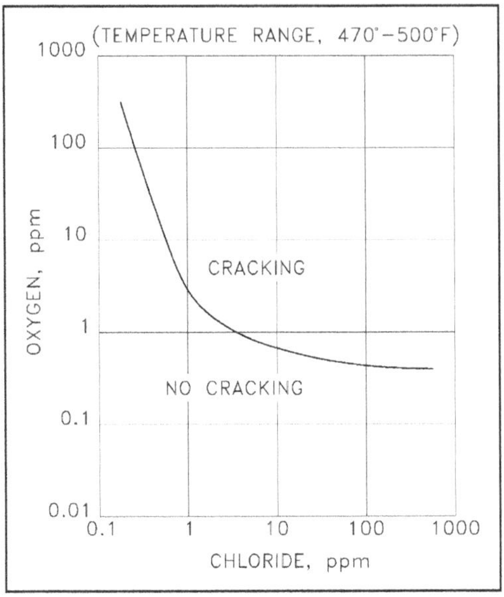

Figure 15 Austenitic Stainless Steel

Caustic Stress Corrosion Cracking

Caustic stress corrosion, or caustic embrittlement, is another form of intergranular corrosion cracking. The mechanism is similar to that of chloride stress corrosion. Mild steels (steels with low carbon and low alloy content) and stainless steels will crack if they are exposed to concentrated caustic (high pH) environments with the metal under a tensile stress. In stress cracking that is induced by a caustic environment, the presence of dissolved oxygen is not necessary for the cracking to occur.

Caustic stress corrosion cracking was first encountered in the operation of riveted steam boilers. These boilers were found to fail on occasion along riveted seams. Failure was attributed to caustic-induced cracking at the highly stressed regions near and under the rivets. Boiler water could easily flow into the crevices which existed under the rivets.

Radiative heating would cause the water in the crevices to boil. As steam was formed, it would escape from the crevice. More boiler water would then flow into the crevice, boil, and pass from the crevice as steam. The net result of this continuing process was concentration of caustic under the rivet. The combination of high stress and high caustic concentrations eventually led to destructive cracking of the boiler vessel.

Where the rate of steam generation (boiling) is high, it is more difficult to eliminate the problem of solute concentration in regions of the boiler. Caustic stress corrosion may concentrate in such regions as the water evaporates rapidly, but sufficient concentration of caustic by such a mechanism to induce stress cracking is considered unlikely.

Available data indicates that caustic concentrations greater than 10,000 ppm, and probably up to 50,000 ppm, are required to induce caustic stress cracking (40,000 ppm NaOH is equivalent to 40 grams per liter or 1 mole per liter). The pH of such a solution is on the order of 14. An alkaline environment is produced and controlled by use of a solution having some properties of a buffer, that is, one that tends to retard or slow a reaction or tends to force it in one direction or the other.

Summary

The important information of this chapter is summarized below.

Specialized Corrosion Summary

- Pitting corrosion occurs where the anodic site becomes fixed in a small area and the formation of holes in an otherwise unaffected area takes place.
- Crevice corrosion is a type of pitting corrosion that occurs specifically within the low flow region of a crevice.
- Stress corrosion cracking (SCC) is a type of intergranular attack corrosion that occurs at the grain boundaries under tensile stress.
- Pitting corrosion requires two conditions to occur, low flow and areas of both high and low oxygen concentration. When these conditions are met a differential aeration cell is established which produces an electron flow from the point of low oxygen concentration to the area of high oxygen concentration. The difference in oxygen concentration is usually due to a low flow condition.
- Pitting corrosion is a hazard due to the possible rapid penetration of the metal with little overall loss of mass. Pitting corrosion is minimized by:

  Avoiding stagnant conditions

  Using the correct metals and alloys that are less susceptible to the corrosion

  Avoiding agents in the medium that cause pitting

  Designing the system and components such that no crevices are present
- Stress corrosion cracking occurs when three conditions are met; the alloy is susceptible to stress corrosion cracking, the alloy is exposed to specific environment, and the alloy is in a stressed condition.
- Chemisorption is the formation of a monomolecular layer of a compound between the metal's surface atoms. This layer separates the metal's atoms thereby weakening the metal and allowing any existing defects to propagate when a stress is applied.

Specialized Corrosion Summary (Cont.)

- Stress corrosion cracking (SCC) is a great concern due to the hazard that it can readily crack metal of appreciable thickness. The crack can lead to a serious failure of the component, or system, and all the attendant results such as:

  Contamination
  Loss of coolant
  Loss of pressure

- SCC is prevented in reactor system by:

  Proper design

  Reducing stress levels

  Avoiding stagnant areas and crevices in heat exchangers, where chlorides and hydroxides might become concentrated.

- Chloride stress corrosion

  The three conditions that must be present for chloride stress corrosion to occur are:

  Chloride ions present
  Dissolved oxygen present
  Metal under tensile stress

  Chloride stress corrosion is controlled by maintaining low chloride ion and oxygen content in the environment and the use of low carbon steels.

- Caustic stress corrosion

  Caustic stress corrosion is similar to chloride stress corrosion. The metals involved are mild steels (steel with low carbon and low alloy content). The metals are exposed to concentrated caustic environments and under tensile stress. The presence of oxygen is not required for cracking to initiate. Concentration of greater than 10,000 ppm is required for cracking to initiate. The level may be higher for different caustic environments.

Intentionally Left Blank.

DOE-HDBK-1015/2-93
JANUARY 1993

DOE FUNDAMENTALS HANDBOOK
CHEMISTRY
Volume 2 of 2

U.S. Department of Energy
Washington, D.C. 20585

FSC-6910

This document has been reproduced directly from the best available copy.

Available to DOE and DOE contractors from the Office of Scientific and Technical Information. P.O. Box 62, Oak Ridge, TN 37831.

Available to the public from the National Technical Information Services, U.S. Department of Commerce, 5285 Port Royal., Springfield, VA 22161.

Order No. DE93012222

ABSTRACT

The *Chemistry* Handbook was developed to assist nuclear facility operating contractors in providing operators, maintenance personnel, and the technical staff with the necessary fundamentals training to ensure a basic understanding of chemistry. The handbook includes information on the atomic structure of matter; chemical bonding; chemical equations; chemical interactions involved with corrosion processes; water chemistry control, including the principles of water treatment; the hazards of chemicals and gases, and basic gaseous diffusion processes. This information will provide personnel with a foundation for understanding the chemical properties of materials and the way these properties can impose limitations on the operation of equipment and systems.

Key Words: Training Material, Atomic Structure of Matter, The Periodic Table of the Elements, Chemical Bonding, Corrosion, Water Chemistry Control, Water Treatment Principles, Chemical Hazards, Gaseous Diffusion Processes

FOREWORD

The *Department of Energy (DOE) Fundamentals Handbooks* consist of ten academic subjects, which include Mathematics; Classical Physics; Thermodynamics, Heat Transfer, and Fluid Flow; Instrumentation and Control; Electrical Science; Material Science; Mechanical Science; Chemistry; Engineering Symbology, Prints, and Drawings; and Nuclear Physics and Reactor Theory. The handbooks are provided as an aid to DOE nuclear facility contractors.

These handbooks were first published as Reactor Operator Fundamentals Manuals in 1985 for use by DOE category A reactors. The subject areas, subject matter content, and level of detail of the Reactor Operator Fundamentals Manuals were determined from several sources. DOE Category A reactor training managers determined which materials should be included, and served as a primary reference in the initial development phase. Training guidelines from the commercial nuclear power industry, results of job and task analyses, and independent input from contractors and operations-oriented personnel were all considered and included to some degree in developing the text material and learning objectives.

The *DOE Fundamentals Handbooks* represent the needs of various DOE nuclear facilities' fundamental training requirements. To increase their applicability to nonreactor nuclear facilities, the Reactor Operator Fundamentals Manual learning objectives were distributed to the Nuclear Facility Training Coordination Program Steering Committee for review and comment. To update their reactor-specific content, DOE Category A reactor training managers also reviewed and commented on the content. On the basis of feedback from these sources, information that applied to two or more DOE nuclear facilities was considered generic and was included. The final draft of each of the handbooks was then reviewed by these two groups. This approach has resulted in revised modular handbooks that contain sufficient detail such that each facility may adjust the content to fit their specific needs.

Each handbook contains an abstract, a foreword, an overview, learning objectives, and text material, and is divided into modules so that content and order may be modified by individual DOE contractors to suit their specific training needs. Each handbook is supported by a separate examination bank with an answer key.

The *DOE Fundamentals Handbooks* have been prepared for the Assistant Secretary for Nuclear Energy, Office of Nuclear Safety Policy and Standards, by the DOE Training Coordination Program. This program is managed by EG&G Idaho, Inc.

OVERVIEW

The *Department of Energy Fundamentals Handbook* entitled *Chemistry* was prepared as an information resource for personnel who are responsible for the operation of the Department's nuclear facilities. An understanding of chemistry will enable contractor personnel to understand the intent of the chemical concerns within their facility. A basic understanding of chemistry is necessary for DOE nuclear facility operators, maintenance personnel, and the technical staff to safely operate and maintain the facility and facility support systems. The information in the handbook is presented to provide a foundation for applying engineering concepts to the job. This knowledge will help personnel understand the impact that their actions may have on the safe and reliable operation of facility components and systems.

The *Chemistry* handbook consists of five modules that are contained in two volumes. The following is a brief description of the information presented in each module of the handbook.

Volume 1 of 2

Module 1 - Fundamentals of Chemistry

Introduces concepts on the atomic structure of matter. Discusses the periodic table and the significance of the information in a periodic table. Explains chemical bonding, the laws of chemistry, and chemical equations.

Appendix A - Basic Separation Theory

Introduces basic separation theory for the gaseous diffusion process. Discusses converter construction and basic operating principals.

Module 2 - Corrosion

Supplies basic information on the chemical interaction taking place during the corrosion process between the environment and the corroding metal.

OVERVIEW (Cont.)

Volume 2 of 2

Module 3 - Reactor Water Chemistry

Describes the chemical measures taken to retard the corrosion often found in water systems. The consequences of radioactivity on facility cooling water systems are also addressed.

Module 4 - Principles of Water Treatment

Details the principles of ion exchange in the context of water purity. Discusses typical water treatment methods and the basis for these methods.

Module 5 - Hazards of Chemicals and Gases

Explains why certain chemicals are considered hazardous to facility personnel. Includes general safety rules on handling and storage.

The information contained in this handbook is by no means all encompassing. An attempt to present the entire subject of chemistry would be impractical. However, the *Chemistry* Handbook does present enough information to provide the reader with a fundamental knowledge level sufficient to understand the advanced theoretical concepts presented in other subject areas, and to better understand basic system and equipment operation.

Department of Energy
Fundamentals Handbook

CHEMISTRY
Module 3
Reactor Water Chemistry

TABLE OF CONTENTS

LIST OF FIGURES

LIST OF TABLES

REFERENCES

- Donald H. Andrews and Richard J. Kokes, Fundamental Chemistry, John Wiley & Sons, Inc., 1963
- Compressed Gas Association, Inc., Handbook of Compressed Gases, 2nd Edition, Reinhold Publishing Corporation, 1981.
- R. A. Day, Jr. and R. C. Johnson, General Chemistry, Prentice Hall, Inc., 1974.
- Dickerson, Gray, Darensbourg and Darensbourg, Chemical Principles, 4th Edition, The Benjamin Cummings Publishing Company, 1984.
- Academic Program for Nuclear Plant Personnel, Volume II, Chemistry, Columbia, MD, General Physics Corporation, Library of Congress Card #A 326517, 1972.
- General Physics Corporation, Fundamentals of Chemistry, General Physics Corporation, 1982.
- Glasstone and Sesonske, Nuclear Reactor Engineering, 3rd Edition, Van Nostrand Reinhold Company, 1981.
- McElroy, Accident Prevention Manual for Industrial Operations Engineering and Technology, Volume 2, 8th Edition, National Safety Council, 1980.
- Sienko and Plane, Chemical Principles and Properties, 2nd Edition, McGraw and Hill, 1974.
- Underwood, Chemistry for Colleges and Schools, 2nd Edition, Edward Arnold, Ltd., 1967.
- Norman V. Steere and Associates, CRC Handbook of Laboratory Safety, 2nd Edition, CRC Press, Inc., 1971.

TERMINAL OBJECTIVE

1.0 Without references, **DESCRIBE** the effects of radiation on reactor water and methods of treatment for the products.

ENABLING OBJECTIVES

1.1 **DESCRIBE** the process of radiolytic decomposition and recombination of water.

1.2 **DESCRIBE** the process of radiolytic decomposition and recombination of nitric acid and ammonia.

1.3 **STATE** the advantage of maintaining excess hydrogen in reactor water.

1.4 **STATE** the three sources of radioactivity in reactor water and each one's decay product.

1.5 **STATE** the following for reactor water chemistry.

a. Nine parameters controlled
b. Reason for controlling each parameter
c. Method of controlling each parameter

1.6 **STATE** the possible effects of abnormal chemistry on core conditions.

Intentionally Left Blank

EFFECTS OF RADIATION ON WATER CHEMISTRY (SYNTHESIS)

Radiation synthesis is a process that takes place in the reactor coolant system. This phenomenon is limited to the reactor coolant system because of the high flux (radiation) levels that exist in the core region and further complicate chemistry control of the reactor plant.

EO 1.1 **DESCRIBE the process of radiolytic decomposition and recombination of water.**

EO 1.2 **DESCRIBE the process of radiolytic decomposition and recombination of nitric acid and ammonia.**

EO 1.3 **STATE the advantage of maintaining excess hydrogen in reactor water.**

EO 1.4 **STATE the three sources of radioactivity in reactor water and each one's decay product.**

Interaction of Radiation

As reactor coolant water passes through the core region of an operating reactor, it is exposed to intense radiation. The major components of the radiation field are neutrons, protons, gamma rays, and high energy electrons (beta particles). These types of radiation interact with the coolant water primarily by an ionization process and have a marked effect on the water itself and on the chemical reactions between substances dissolved in the water. This section discusses these effects, and in particular the effects that involve gases dissolved in reactor coolant.

The interaction of radiation with matter produces ion pairs. Usually, the negative member of the ion pair is a free electron and the positive member is a polyatomic cation, the exact nature of which depends on the particular substance being irradiated. For example, the interaction of radiation with water is illustrated by the following reaction.

$$H_2O + \text{radiation} \rightarrow e^- + H_2O^+ \qquad (3\text{-}1)$$

Both of these species are very reactive chemically, and there are several reaction pathways available to each. Some of these mechanisms are very complex and are usually of little practical value to the reactor operator, who is more concerned with the overall, observable effects. In the case of water, the overall effect of irradiation is shown in the following reaction.

$$2H_2O + \text{radiation} \rightarrow 2H_2 + O_2 \qquad (3\text{-}2)$$

Because this result is not at all apparent from Reaction (3-1), the following section describes the intermediate processes in some detail. This discussion is presented only to illustrate the types of reaction mechanisms that occur in irradiated solutions. Subsequent discussions primarily involve only the overall effects of these processes.

Reaction (3-1) shows that irradiation of pure water produces an electron and a H_2O^+ ion. As stated, both species are highly reactive. The H_2O^+ ion rapidly reacts with a water molecule as follows.

$$H_2O^+ + H_2O \rightarrow H_3O^+ + OH \qquad (3\text{-}3)$$

The species OH is an uncharged hydroxyl group. Neutral groups such as this, in which all chemical bonding capacity is not satisfied, are common intermediate species in chemical reactions and are called radicals or sometimes free radicals.

The electron produced by Reaction (3-1) first forms a species called the hydrated electron, denoted by e_{aq}^-. The hydrated electron may be thought of as resulting from the interaction of a negative electron with the positive end of a polar water molecule. This is analogous to the formation of hydronium ions by interaction of a positive proton (H^+) with the negative end of a water molecule. Because the water molecules associated with hydrated electrons do not participate in subsequent chemical reactions, they are not shown in chemical equations, and the hydrated electron (e_{aq}^-) is used instead.

Hydrated electrons may interact with H_3O^+ ions in solution or with water molecules. Both reactions produce another reactive species, atomic hydrogen.

$$e_{aq}^- + H_3O^+ \rightarrow H + H_2O \qquad (3\text{-}4)$$

or

$$e_{aq} + H_2O \rightarrow H + OH \qquad (3\text{-}5)$$

Reaction (3-4) usually predominates.

Because Reactions (3-4) and (3-5) are slow compared to that in Reaction (3-3), there are three reactive species present at any one time: hydroxyl radicals (OH), hydrated electrons (e_{aq}^-), and hydrogen atoms (H). These species may undergo any of several possible reactions such as the following.

$$OH + OH \rightarrow H_2O_2 \text{ (hydrogen peroxide)} \qquad (3\text{-}6)$$

$$OH + H \rightarrow H_2O \qquad (3\text{-}7)$$

$$H + H \rightarrow H_2 \qquad (3\text{-}8)$$

$$H + e_{aq}^- \xrightarrow{H_2O} H_2 + OH^- \qquad (3\text{-}9)$$

$$H_2 + OH \rightarrow H_2O + H \qquad (3\text{-}10)$$

Hydrogen peroxide, formed by Reaction (3-6), may also react with the original reactive species, but at high temperatures H_2O_2 is unstable, and the predominant reaction is decomposition.

$$2H_2O_2 \rightarrow O_2 + 2H_2O \qquad (3\text{-}11)$$

To illustrate the overall result of these reactions, let us assume that each of the reactive species produced by successive steps in the irradiation of water reacts in only one way. That is, whenever several reactions of a particular substance are possible, assume that one predominates to such an extent that the others are negligible. The following set of reactions is one possibility. In some cases, entire reactions are multiplied by a factor to allow cancellation of terms when the reactions are summed.

$$4(H_2O + \text{radiation} \rightarrow e^- + H_2O^+) \qquad (3\text{-}1)$$

$$4(H_2O^+ + H_2O \rightarrow H_3O^+ + OH) \qquad (3\text{-}3)$$

$$e_{aq} + H_3O^+ \rightarrow H + H_2O \qquad (3\text{-}4)$$

$$2(OH + OH \rightarrow H_2O_2) \qquad (3\text{-}6)$$

$$2(H + H \rightarrow H_2) \quad (3\text{-}8)$$

$$2H_2O_2 \rightarrow O_2 + 2H_2O \quad (3\text{-}11)$$

Net reaction: $$8H_2O + \text{radiation} \rightarrow 2H_2 + O_2 + 6H_2O$$

or

$$2H_2O + \text{radiation} \rightarrow 2H_2 + O_2 \quad (3\text{-}12)$$

The net result of these reactions is simply the decomposition of water. If H_2 and O_2 are allowed to escape from solution as gases, the reaction continues as written. If, however, the water is contained in a closed system under pressure (as in a reactor coolant system), H_2 and O_2 are confined, and an equilibrium state is reached because radiation also causes the reverse of Reaction (3-2) to take place. Primarily neutron and gamma radiation induce both the decomposition of water and the recombination of H_2 and O_2 to form water. Thus, it is appropriate to write Reaction (3-2) as a radiation-induced equilibrium reaction.

$$2H_2O \underset{\text{radiation}}{\overset{\text{radiation}}{\rightleftharpoons}} 2H_2 + O_2 \quad (3\text{-}12)$$

To arrive at the overall effect of radiation on water, the above process involved the assumption that only one reaction pathway is available to each reactive species. This was done primarily for convenience because inclusion of every possible reaction in the summation process becomes rather cumbersome. Even if all the reactions are taken into account, the net result is the same as Reaction (3-12), which is reasonable because inspection of Reactions (3-3) through (3-11) shows that the only stable products are H_2, O_2, and H_2O (H_3O^+ and OH^- combine to form water, and H_2O_2 decomposes at high temperature). Perhaps not as obvious, more water is consumed than is produced in these reactions, and the net result is the initial decomposition of water that proceeds until equilibrium concentrations of H_2 and O_2 are established.

Before discussing the effects of radiation on other processes, chemical equilibrium in the presence of ionizing radiation should be mentioned. Equilibrium processes in aqueous solutions are discussed briefly in Module 1, which states that temperature influences the equilibrium. Ionizing radiation also influences the equilibrium of these solutions.

Radiation has an effect on the equilibrium in the case of water. In the absence of radiation, water does not spontaneously decompose at 500°F and the equilibrium lies far to the right.

$$2H_2 + O_2 \leftrightarrows 2H_2O$$

When irradiated, however, water does decompose, as shown above. Also, H_2 and O_2 do not normally react at 500°F because a large activation energy is required to make the reaction occur. Radiation, in effect, supplies this activation energy, and the reaction takes place readily. Thus, radiation increases the rates of both forward and reverse reactions, although not by the same factor.

In general, the effect of radiation on the equilibrium for a given reaction cannot be predicted quantitatively. The situation is further complicated by the observation that the effect on the equilibrium may vary with the intensity of the radiation. In nuclear facilities, the effect may vary with the power level of the facility. In most cases, this complication is not a severe problem because the direction of the effect is the same; only the degree or magnitude of the effect varies with the intensity of the radiation.

As noted several times previously, reactor coolant is maintained at a basic pH (in facilities other than those with aluminum components or those that use chemical shim reactivity control) to reduce corrosion processes. It is also important to exclude dissolved oxygen from reactor coolant for the same reason. As shown in the preceding section, however, a natural consequence of exposing pure water to ionizing radiation is production of both hydrogen and oxygen. The addition of a base to control pH has essentially no effect on this feature.

To prevent the formation of oxygen in reactor coolant, hydrogen is added. Hydrogen suppresses the formation of oxygen primarily by its effect on the reactions that OH radicals, produced by Reaction (3-3), undergo. In the presence of excess hydrogen, hydroxyl radicals react predominantly by Reaction (3-10) rather than as in Reactions (3-6) through (3-8).

$$H_2 + OH \rightarrow H_2O + H \qquad (3\text{-}10)$$

Hydrogen atoms from this equation subsequently react to form H_2 and H_2O by Reactions (3-7), (3-8), and (3-9). None of these reactions leads to O_2, or H_2O_2, which decomposes to form O_2 and H_2O at high temperatures. Thus, the addition of H_2 to reactor coolant largely eliminates production of free oxygen.

Another way of viewing the effect of hydrogen on reactor coolant is through its effect on the equilibrium of the reaction.

$$2H_2O + \text{radiation} \rightarrow 2H_2 + O_2 \quad (3\text{-}12)$$

By LeChatelier's principle, the addition of excess hydrogen forces the equilibrium to the left, which requires that O_2 be consumed. If the dissolved hydrogen concentration is sufficiently large, only a very small amount of oxygen will be present at equilibrium. Normally, therefore, reactor coolant contains excess dissolved hydrogen, and there is no significant net decomposition of water by radiation.

Reactor coolant makeup water usually contains a small amount of air, which is composed primarily of nitrogen and oxygen in a volume ratio of 4:1 (80 percent nitrogen, 20 percent oxygen). These gases undergo radiation-induced reactions. The reactions are the same as those that occur in certain accident situations and are included in the following discussion.

In addition to the small amount of air normally dissolved in makeup water, there is a small possibility that air may be accidentally injected directly into the reactor coolant system.

Whenever air enters the reactor coolant system, and the reactor is operating, the most immediate reaction involves oxygen from the air and hydrogen, which is normally present in the coolant.

$$2H_2 + O_2 \underset{\text{radiation}}{\overset{\text{radiation}}{\rightleftharpoons}} 2H_2O \quad (3\text{-}13)$$

That is, the addition of O_2 disturbs the above equilibrium and causes the equilibrium to shift to the right, consuming both H_2 and O_2. The concentration of hydrogen normally maintained in reactor coolant is such that small amounts of oxygen will be rapidly consumed before any excess oxygen can cause severe corrosion problems to occur.

Reaction (3-13) also consumes oxygen added to the reactor coolant as a natural consequence of air dissolved in makeup water. Other than initial fill of the reactor coolant system, the situations that require the largest amounts of makeup water are feed and bleed operations to correct an abnormal chemistry parameter or cooldown after some period of reactor operation. In this case, gamma radiation from the decay of fission products in the reactor core continues to induce the H_2 - O_2 reaction for some period after shutdown. During initial fill and long shutdown periods, chemicals other than hydrogen (e.g. hydrazine) may be added to reactor coolant to remove any dissolved oxygen.

After essentially all of the oxygen has been consumed by reaction with hydrogen, the nitrogen contained in air will remain. For small air additions, some hydrogen will also remain; thus, the reactor coolant will contain both dissolved hydrogen and dissolved nitrogen. These two gases do not react in an unirradiated solution at low temperature and pressure. When exposed to radiation, however, the gases do react by the following reaction.

$$3H_2 + N_2 \overset{\text{radiation}}{\rightleftharpoons} 2NH_3 \text{ (ammonia)} \tag{3-14}$$

Again, this is an equilibrium reaction, and radiation induces the reaction in both directions. Ammonia (NH_3) produced by this reaction combines with water to form ammonium hydroxide (NH_4OH).

$$NH_3 + H_2O \rightleftharpoons NH_4^+ + OH^- \tag{3-15}$$

Under the operating conditions of reactor coolant, Reaction (3-14) is far from complete. In most cases, less than about 10 percent of the nitrogen will be converted to ammonia. If no additional base were added to reactor coolant, Reaction (3-14) would be sufficient to cause the coolant to be mildly basic, pH 9. In the presence of added base, however, the reaction has only a very slight and negligible effect on pH.

If the base NH_3 were used to control reactor coolant pH, the reverse of Reaction (3-14) would be more important. The reverse step of this reaction requires that some of the ammonia added to the coolant decompose into N_2 and H_2. Because operating conditions favor this step of the equilibrium, rather than formation of NH_3, it would be expected that most of the ammonia added would decompose. However, the rate of the ammonia decomposition reaction is slow, and the pH of reactor coolant can be maintained in the required range. It should also be noted that the decomposition of NH_3 would produce hydrogen gas in significant concentrations in reactor coolant (sufficient to satisfy normal H_2 requirements).

In the event that a large quantity of air is injected into the reactor coolant system, the inventory of dissolved hydrogen would be rapidly depleted by Reaction (3-13). If the amount of air injected is sufficiently large, there could be oxygen remaining in the coolant after depletion of the hydrogen. In this case, another reaction is available to the oxygen and nitrogen in the air.

$$2N_2 + 5O_2 + 2H_2O \overset{\text{radiation}}{\rightleftharpoons} 4HNO_3 \tag{3-16}$$

Nitric acid (HNO_3) produced by this reaction will neutralize any base contained in the coolant, and if sufficient acid is produced, the coolant will acquire an acidic pH.

Normally, the amount of hydrogen maintained in the reactor coolant, in conjunction with other precautions employed, greatly reduces the probability that the amount of oxygen entering the coolant will be sufficient to lead to Reaction (3-16). If a large amount of air were accidentally added to the reactor coolant, one solution would be to add more hydrogen. The added hydrogen would react with remaining oxygen, disrupting the equilibrium of Reaction (3-16) causing the reverse step of that reaction to occur. When all the oxygen has been removed, H_2 and N_2 could react by Reaction (3-14) and help reestablish a basic pH. The relationship between these reactions and pH following the initial oxygen addition, and a subsequent hydrogen addition, is illustrated in Figure 1.

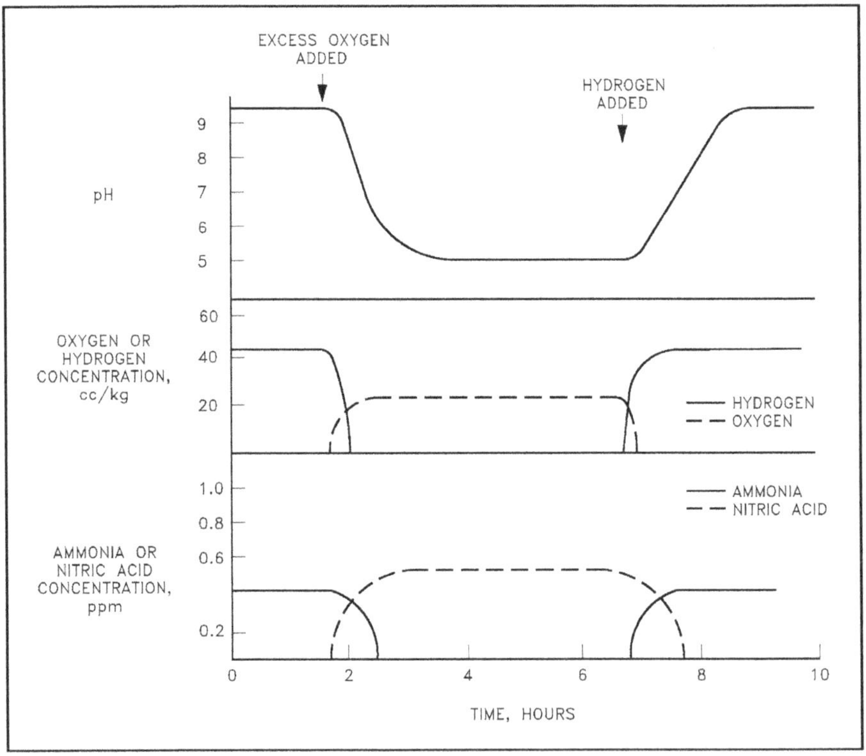

Figure 1 Change in pH, Gas Concentration, and Nitrogen Compounds With Excess Oxygen Added

In the preceding discussion, the reactions possible after the addition of air to reactor coolant containing hydrogen were described. These are Reactions (3-13), (3-14), and (3-16). The relative rates of these reactions are of considerable importance. Briefly, Reaction (3-13) is much more rapid than either of the others, and Reaction (3-16) is faster than Reaction (3-14). Thus, the sequence of reactions is as described in the preceding sections. H_2 and O_2 react to form water. If hydrogen remains, it undergoes an incomplete reaction with N_2 to form small amounts of NH_3. If O_2 remains after all the H_2 has been consumed, the O_2 - N_2 reaction produces nitric acid.

The flux of neutrons and protons in a nuclear reactor core region leads to several important nuclear reactions with the constituent atoms of water. Most of these reactions involve oxygen isotopes and fast neutrons or protons.

In many cases, the absorption of a fast neutron by a nucleus is immediately followed by ejection of a proton. These reactions are called neutron-proton or n-p reactions and are commonly written (using the ^{16}O reaction to illustrate) in the following manner.

$$^{16}_{8}O\ (n, p)\ ^{16}_{7}N\ (t_{1/2} = 7.13\ \text{seconds}) \quad (3\text{-}17)$$

In this notation, the original isotope that undergoes the reaction is written first, the product isotope is last, and the two are separated by, in order, the particle absorbed and the particle emitted. The isotope $^{16}_{7}N$ decays to $^{16}_{8}O$ with a 7.13-second half-life by emitting a beta particle (β^-) and a high-energy gamma ray (6 Mev predominantly).

$$^{16}_{7}N \rightarrow {}^{16}_{8}O + \beta^- + \gamma$$

Oxygen-17 undergoes a similar reaction.

$$^{17}_{8}O\ (n, p)\ ^{17}_{7}N\ (t_{1/2} = 4.1\ \text{seconds}) \quad (3\text{-}18)$$

The isotope $^{17}_{7}N$ decays by emission of a beta particle, a neutron, and a gamma ray.

$$^{17}_{7}N \rightarrow {}^{16}_{8}O + \beta^- + {}^{1}_{0}n + \gamma$$

Reactions (3-17) and (3-18) have no significant chemical effect on reactor coolant because of the relatively small number of atoms that undergo these reactions. They are of considerable importance, however, because the radioactive species $^{16}_{7}N$ and $^{17}_{7}N$ are carried outside the core region by the flow of reactor coolant. The neutrons and high-energy gamma rays emitted by these isotopes easily penetrate the piping and components that contain the coolant and are important considerations in the design of shielding for nuclear facilities. Because the half-lives of these isotopes are very short, they decay to low levels very rapidly after shutdown and are, therefore, of little concern during such periods.

Two other nuclear reactions with oxygen isotopes are shown below.

$$^{18}_{8}O\ (p, n)\ ^{18}_{9}F\ (t_{1/2} = 112 \text{ minutes})$$

$$^{16}_{8}O\ (p, \alpha)\ ^{13}_{7}N\ (t_{1/2} = 10 \text{ minutes})$$

An ejected alpha particle is indicated by α. The protons that cause these reactions result from inelastic collisions of fast neutrons with hydrogen atoms in water molecules. The radioactivity levels of these isotopes are much lower than the levels of $^{16}_{7}N$ and $^{17}_{7}N$ during reactor facility operation. However, during the period from a few minutes to about five hours after reactor shutdown or after removing a coolant sample from the system, $^{13}_{7}N$ and $^{18}_{9}F$ are the principal sources of radioactivity in the reactor coolant of most reactor facilities.

$$^{13}_{7}N \rightarrow\ ^{13}_{6}C +\ ^{0}_{+1}\beta$$

$$^{18}_{9}F \rightarrow\ ^{18}_{8}O +\ ^{0}_{+1}\beta$$

The only significant nuclear reaction that occurs with hydrogen involves deuterium ($^{2}_{1}H$), which comprises about 0.015 percent of natural hydrogen.

$$^{2}_{1}H\ (n, \gamma)\ ^{3}_{1}H\ (t_{1/2} = 12.3 \text{ years})$$

Tritium ($^{3}_{1}H$) decays by emission of a very weak β particle (0.02 Mev) and no gamma rays. Thus, tritium is not a radiological hazard unless it enters the body in significant amounts. Tritium can enter the body through inhalation or ingestion. It is also possible to absorb forms of tritium through the skin.

Summary

The important information in this chapter is summarized below.

Effects of Radiation on Water Chemistry (Synthesis) Summary

- The intense radiation inside the core of an operating nuclear reactor produces several chemical effects on the reactor coolant itself and on substances dissolved in the coolant. Radiation causes pure water to decompose into H_2 and O_2. The decomposition is suppressed by adding excess hydrogen.

$$2H_2O \rightleftharpoons 2H_2 + O_2$$

- Excess hydrogen is added to suppress the decomposition of reactor water. It also reacts with any oxygen that enters the reactor coolant system, usually as a component of air in makeup water, provided the amount of oxygen is not excessive. If the amount of oxygen is more than enough to deplete the hydrogen, the excess oxygen reacts with nitrogen (also a component of air) and forms nitric acid. In the case of addition of very large amounts of air, the amount of nitric acid produced may be more than enough to neutralize the normally basic coolant and cause it to become acidic.

- Radiation induces the combination of N_2 and H_2 to form ammonia, although the extent of this reaction is small and usually has a negligible effect on the pH of reactor coolant. All of the reactions in this chapter are reversible and reach an equilibrium state under normal operating conditions. Changes in the concentrations of any of the reactants disturb the equilibrium and causes the reaction to shift in the direction which restores the equilibrium.

$$N_2 + 3H_2 \rightleftharpoons 2NH_3$$

- Radiation also produces several nuclear reactions in reactor coolant. The products ${}^{16}_{7}N$ and ${}^{17}_{7}N$, of two of these reactions, contribute radioactivity to the reactor coolant, adding significantly to the shielding requirements for nuclear reactors. Others, ${}^{13}_{7}N$ and ${}^{18}_{9}F$, are also major contributors to the radioactivity in reactor coolant.

$${}^{16}_{7}N \rightarrow {}^{16}_{8}O + \beta^- + \gamma \qquad {}^{17}_{7}N \rightarrow {}^{16}_{8}O + \beta^- + {}^{1}_{0}n + \gamma$$

$${}^{13}_{7}N \rightarrow {}^{13}_{6}C + {}^{0}_{+1}\beta \qquad {}^{18}_{9}F \rightarrow {}^{18}_{8}O + {}^{0}_{+1}\beta$$

CHEMISTRY PARAMETERS

The reasons for control of selected chemistry parameters, and some of the more common methods of controlling these parameters will be discussed. No attempt will be made to identify specific values of any of the parameters discussed because of the number of different reactor facilities involved, but an overview concerning the bases and common methods used will be included. For operating values and specifications, users should refer to local facility publications. In addition, some information on tritium is provided.

EO 1.5 STATE the following for reactor water chemistry.

- **a. Nine parameters controlled**
- **b. Reason for controlling each parameter**
- **c. Method of controlling each parameter**

EO 1.6 STATE the possible effects of abnormal chemistry on core conditions.

Specific Parameters

Specific chemical parameters vary from facility to facility but generally include the following: pH, dissolved oxygen, hydrogen, total gas content, conductivity, chlorides, fluorine, boron, and radioactivity. For the parameters indicated, control is generally achieved by one or more of three basic processes.

(1) Ion exchange in the primary system demineralizer(s) or by supplemental chemical additions

(2) Oxygen scavenging by hydrogen or hydrazine addition

(3) Degassification

Table 1 lists the more common chemistry parameters measured and/or controlled, the reasons each is measured and/or controlled, and control methods utilized.

TABLE 1
Summary of Reactor Coolant Chemistry Control

Parameter	Reason	Method of Control
pH	• To inhibit corrosion • To protect corrosion film • To preclude caustic stress corrosion	• Ion exchange • Ammonium hydroxide addition • Nitric acid addition
Dissolved Oxygen	• To inhibit corrosion	• Hydrogen addition • Hydrazine addition
Hydrogen	• To scavenge oxygen • To suppress radiolytic decomposition of water • To scavenge nitrogen • To preclude hydrogen embrittlement	• Hydrogen addition • Degassification
Total Gas Content	• To protect pumps • To indicate air in leakage	• Degassification • Deaeration of makeup water
Conductivity	• To minimize scale formation • To indicate increased corrosion	• Ion exchange • Feed and Bleed
Chlorides	• To preclude chloride stress corrosion	• Ion exchange • Feed and bleed
Fluorine	• To preclude corrosion of Zr cladding	• Ion exchange • Feed and Bleed
Boron	• To control reactivity	• Boric acid addition
Radioactivity	• To indicate increased corrosion • To indicate a crud burst • To indicate a core fuel defect • To monitor effectiveness of demineralizer	• Ion exchange • Feed and bleed

pH

The reason for controlling pH in the reactor coolant system is to minimize and control corrosion. As discussed in Module 1, the presence of excess H^+ ions in solution results in an acidic condition. In reactor facilities (except those containing aluminum components), acidic conditions are detrimental to the materials of construction in a number of ways. An acidic condition in the primary coolant results in processes that are potentially harmful to the system as follows. First, a low pH promotes rapid corrosion by deteriorating or "stripping off" the protective corrosion film, and second, corrosion products such as ferrous oxide (Fe_3O_4), which is predominant in the corrosion film, are highly soluble in an acidic solution. Figure 2 shows how the corrosion rate increases as the pH decreases. Thus for facilities not using aluminum components, a neutral or highly basic pH is less corrosive.

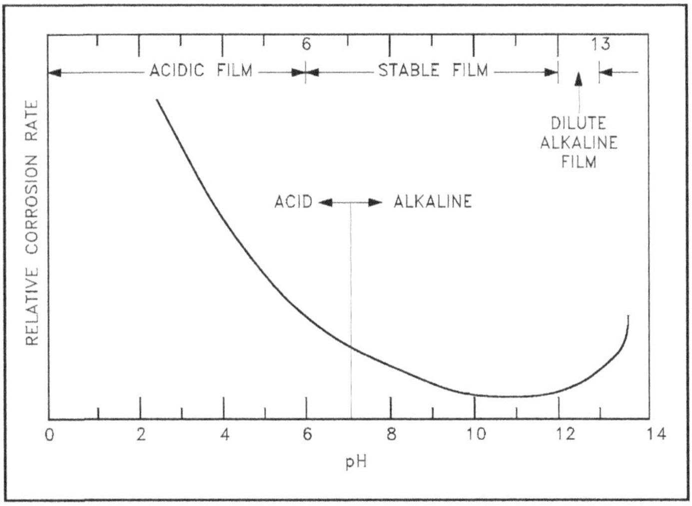

Figure 2 Corrosion Rate vs. pH for Iron

In nuclear facilities that do not use chemical shim to control reactivity, pH is normally maintained at a relatively high value, such as a pH of about 10. In these facilities the upper limit for pH is set based on caustic stress corrosion considerations because caustic stress corrosion becomes more probable as higher pH values are approached.

In facilities that use chemical shim reactivity control (chemical shim involves the addition of boron in the form of boric acid) the pH is maintained at a much lower value. A low pH is necessary because of the large amounts of boric acid added to the reactor coolant. Accordingly, pH in these facilities is maintained as high as possible consistent with the reactivity requirements of the nuclear facility, with pH range from 5 to 7 being common.

In facilities using aluminum components, pH is maintained on the acidic side of the scale because of the corrosion characteristics of aluminum discussed in Module 2. In these facilities pH may be controlled by the addition of a dilute nitric acid (HNO_3) solution to the reactor coolant system in conjunction with an ion exchange system of some type.

Regardless of the pH range maintained, most facilities use an ion exchange process (described in Module 4) to help control pH. For the high pH facilities, the most common means of control is the use of a lithium or an ammonium form cation and a hydroxyl form anion. When lithium is used, it must be ^7Li because other lithium isotopes produce tritium, which represents a significant biological hazard to personnel. In facilities that employ high pH chemistry control and do not use chemical shim reactivity control, it is sometimes necessary to add a strong base solution such as ammonium or lithium hydroxide. When chemical additions are used for pH control, facility design and operating procedures are utilized to preclude overconcentration at any point in the system, which may lead to caustic stress corrosion conditions. Many reactions that take place in the reactor coolant system can affect pH; therefore chemistry control must be considered carefully to preclude upsetting the pH balance provided by the ion exchanger.

Dissolved Oxygen

Control of the dissolved oxygen content in the reactor facility system is of paramount importance because of its contribution to increased corrosion. The base reactions of concern regarding high concentrations of dissolved oxygen are the following.

$$3Fe + 2O_2 \rightarrow Fe_3O_4 \qquad (3\text{-}19)$$

$$4Fe + 3O_2 \rightarrow 2Fe_2O_3 \qquad (3\text{-}20)$$

They are dependent on both the concentration of oxygen and temperature. Reaction (3-19) is predominant at high temperatures (>400°F) in the presence of lower oxygen concentrations. This corrosion film, ferrous oxide, is also known as magnetite and is a black, generally tightly-adherent film that provides a protective function to surfaces within the facility. Reaction (3-20) occurs at temperatures below about 400°F in the presence of higher oxygen concentrations. Ferric oxide (Fe_2O_3) is more commonly known as rust and is generally a reddish color. This corrosion product adheres loosely to surfaces and is therefore easily removed and transported throughout the system for subsequent deposition and possible irradiation. In either of the reactions, the corrosion rate is accelerated by increased concentrations of dissolved O_2 and can be aggravated further by the presence of other substances that may be present in the system.

In addition to the direct contribution to corrosion, oxygen reacts with nitrogen to lower the pH of the reactor water, which also results in an increased rate of corrosion. Oxygen and nitrogen react to form nitric acid by the following reaction.

$$2N_2 + 5O_2 + 2H_2O \xrightarrow{\text{radiation}} 4HNO_3 \text{ (nitric acid)}$$

In all the reactions presented, it can be seen that oxygen concentrations promote corrosion. It follows then that if corrosion is to be minimized, oxygen concentrations must be maintained as low as possible. In most nuclear facility reactor coolant systems, the limit for dissolved oxygen concentrations is expressed in ppb (parts per billion). Concentration may be monitored on a continuous basis by using an in-line analyzing system or periodically by withdrawing a sample volume and analyzing that sample. Monitoring oxygen levels is done not only to ensure that no oxygen is available for corrosion, but also to indicate the introduction of air into the system.

Hydrogen

Because the presence of dissolved oxygen contributes to most mechanisms of corrosion, the concentration of oxygen is controlled and reduced by the addition of scavenging agents in most facilities. Hydrogen gas (H_2) and hydrazine (N_2H_4) are the scavenging agents normally used to eliminate dissolved oxygen from the reactor coolant system. These substances scavenge oxygen by the following reactions.

$$2H_2 + O_2 \underset{\text{radiation}}{\overset{\text{radiation}}{\rightleftharpoons}} 2H_2O \quad (3\text{-}13)$$

$$N_2H_4 + O_2 \rightarrow 2H_2O + N_2 \quad (3\text{-}21)$$

Because hydrazine decomposes rapidly at temperatures above about 200°F (forming NH_3, H_2, and N_2), hydrogen gas is used as the scavenging agent during hot operation and hydrazine is used when the reactor coolant system is cooled below 200°F.

$$2N_2H_4 \xrightarrow{\text{(Heat)}} 2NH_3 + N_2 + H_2$$

(decomposition of hydrazine)

The decomposition reactions of hydrazine pose additional problems in chemistry control. Even if sufficient hydrazine were added to overcome the loss due to decomposition, instability of coolant pH would probably occur by the following reactions.

$$2N_2 + 5O_2 + 2H_2O \rightarrow 4HNO_3 \text{ (acid)} \quad (3\text{-}16)$$

$$3H_2 + N_2 + 2H_2O \rightarrow 2NH_4OH \text{ (base)} \quad (3\text{-}22)$$

The use of hydrogen gas at temperatures above 200°F precludes the generation of the compounds formed by Reactions (3-16) and (3-22). In addition, hydrogen is compatible with the high flux levels present in the reactor core. Accordingly, advantage may be taken of the reversibility of the radiolytic decomposition of water. The following reaction illustrates the scavenging process utilizing hydrogen.

$$2H_2 + O_2 \underset{\text{radiation}}{\overset{\text{radiation}}{\rightleftarrows}} 2H_2O \qquad (3\text{-}13)$$

As indicated, the reaction is an equilibrium process and will therefore depend on the relative concentrations of the reactants and the products. By maintaining an excess of hydrogen (H_2), the reaction is forced to shift to the right and theoretically eliminates any dissolved oxygen that may be present. As long as an inventory of H_2 is present in the coolant, dissolved oxygen will be eliminated or forced to recombine immediately after radiolytic decomposition, thereby being unavailable for corrosion reactions.

A boiling water reactor (BWR) facility is susceptible to corrosion, resulting from dissolved oxygen, in the same reactions as are present in a pressurized water reactor (PWR). However, because of the design of these facilities the use of chemical additives is prohibited because continuous concentration would occur in the reactor vessel due to boiling. Boiling would result in a plating out process, and the irradiation of these concentrated additives or impurities would create an extreme environment of radiation levels as well as adverse corrosion locations.

By the very nature of operation of a BWR facility, the buildup of high concentrations of dissolved oxygen is prevented. Because boiling is occurring in the reactor vessel and the steam generated is used in various processes and subsequently condensed, removal of dissolved gases is a continual process. As stated, boiling is an effective means of removing gases from a solution. If we were to compare the oxygen content of the steam and the water in a BWR, we would find typical concentrations of 100 ppb to 300 ppb in the water and 10,000 ppb to 30,000 ppb in the steam. This concentration process is continuous during operation, and the dissolved oxygen remains in the gaseous state and is subsequently removed in the condensing units along with other noncondensible gases. As with PWR facilities, BWR facilities minimize the introduction of dissolved oxygen by pretreating makeup water by some method. The large oxygen concentrations measured in the steam system result primarily from the radiolysis of water according to Reaction (3-12), and as operation is continued, the equilibrium concentration of 100 ppb to 300 ppb is established. This concentration of oxygen is consistent with the objective of minimizing corrosion.

Total Gas

Total gas concentration in the reactor coolant system is another parameter of concern. Total gas is the sum of all gases contained in the coolant system and is made up primarily of hydrogen (H_2), nitrogen (N_2), argon (Ar), and oxygen (O_2). The small amounts of fission gases (Kr and Xe) normally present in the system may also contribute to the total gas concentration; however, under normal conditions these are essentially undetectable. Total gas is of concern because high concentrations can result in the formation of gas pockets in areas that are high points of the system where low or stagnant flow conditions exist. Of particular concern in PWR facilities are the erosion and corrosion that may occur on the impellers of the primary coolant pumps. As the concentration of gas is increased, the probability of the gas coming out of solution in significant amounts in areas of low pressure is also increased. This low pressure condition exists at the inlet to the primary coolant pump impeller (where centrifugal pumps are utilized). As these gas bubbles are forced back into solution on the high pressure side of the impeller, erosion can occur as a result of the gas bubble impinging on the impeller. In extreme concentrations of total gas, loss of pump priming and cavitation can occur with resultant mechanical pump damage.

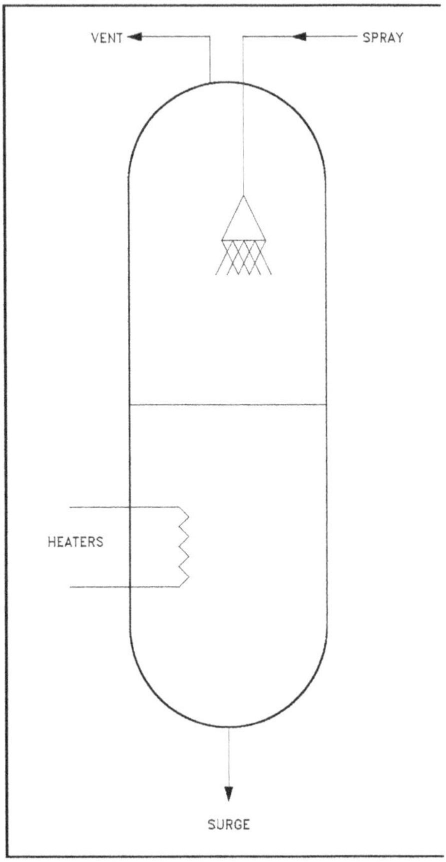

Figure 3 Pressurizer

Reduction of total gas concentrations in PWRs is normally accomplished by the venting of a steam space. In those facilities utilizing a pressurizer, the steam space in the top of the pressurizer is designed to accomplish this venting operation either continuously or intermittently. This process of reducing the total gas concentration is generally referred to as degassification. A typical PWR pressurizer with degassification piping is shown in Figure 3.

Degassification is not normally required in a BWR because of its design. As discussed previously, the boiling action in the reactor vessel strips dissolved gases from the water, and they are continuously removed in the condensing phase of the energy cycle.

Conductivity

Conductivity of reactor facility water is measured to provide an indication of dissolved ionic substances in the coolant. Conductivity measurements provide quantitative rather than qualitative information because it is possible to determine the total conductivity of the ions present, but not the specific types of ions present. Because many ions such as iron (Fe^{+++}), chromium (Cr^{+++}), copper (Cu^{++}) and aluminum (Al^{+++}) are susceptible to forming oxides and plating out as scale on heat transfer surfaces, reactor coolant conductivity is normally controlled at a level as low as practicable and consistent with pH. By monitoring conductivity levels in the reactor facility systems, the operator is able to cross check the chemistry of these systems, thereby achieving a higher confidence level in the parameters measured.

Regardless of the operating limits specified for a given reactor facility, operating relationships can be established between pH and conductivity levels of the coolant. Figure 4 shows a typical relationship of the pH and conductivity of a reactor coolant system using high pH, ammonium hydroxide chemistry control as a function of the ammonia (NH_3) concentration.

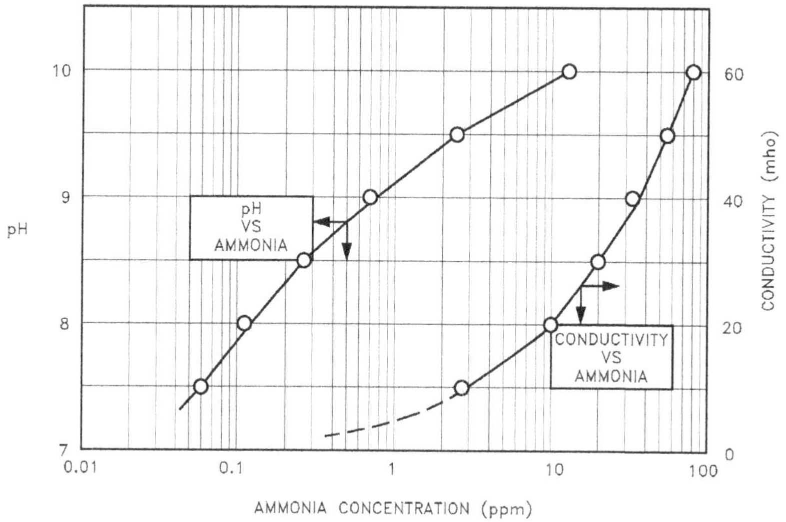

Figure 4 pH and Conductivity as a Function of NH_3 Concentration

Figure 5 shows the theoretical relationship of pH versus conductivity in a solution containing pure water. A second curve is added to the graph that illustrates the relationship that exists when nitric acid (HNO_3) is used as a pH control agent (such as may be utilized in facilities with aluminum components).

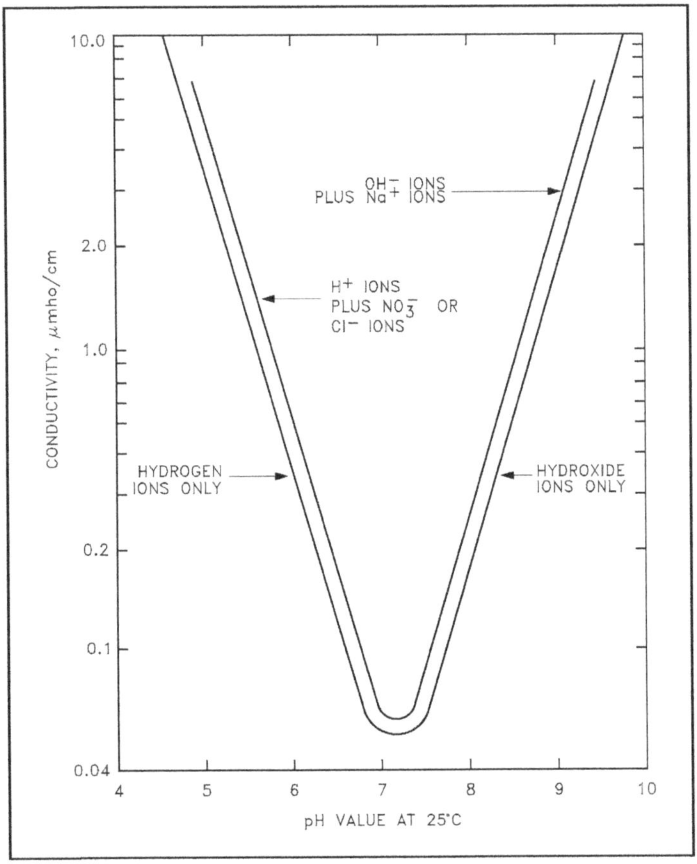

Figure 5 Theoretical Conductivity as a Function of pH

In both Figure 4 and Figure 5, a definite relationship exists between pH and conductivity, assuming no foreign ions are present. A similar graph could be constructed for those facilities using cation resins of a different base such as lithium or barium.

The key point of this discussion is the realization that a theoretical, or calculated, relationship does exist, and measurements that vary appreciably from the theoretical values should be investigated and corrective action taken.

Excessively high conductivity levels are an indication of the presence of undesired ions. This condition warrants further investigation to locate the source of the impurity because, in addition to other chemistry problems, it contributes to general corrosion by increasing the reaction rates of the electrochemical cells. The purity of the makeup water, and any pH control agents added, should be verified to determine the cause. pH should also be checked because of the relationship of these parameters. Other chemistry parameters should also be checked, such as Cl^- and F^-. After the cause of high conductivity has been determined, appropriate steps should be taken to return conductivity to its normal value. One method that is often used is a feed and bleed procedure whereby water is added to and drained from the facility at the same time. If this method is used, verification of makeup water purity must be ensured to prevent compounding the problem.

Low conductivity is also an indicator of a potential problem because, in high purity basic systems, the only possible cause of low conductivity is a low pH. For example, in a system using high pH ammonium hydroxide control, the introduction of air into the facility could result in the formation of nitric acid (HNO_3) with a reduction in pH by the following reaction.

$$2N_2 + 5O_2 + 2H_2O \leftrightarrows 4HNO_3 \qquad (3\text{-}16)$$

Conductivity decreases even more than would be expected because of the formation of NH_4NO_3. NO_3^- is not as conductive as OH^-, so the NH_4NO_3 results in a lower conductivity than NH_4OH. This neutralization of NH_4OH is shown by the following reaction.

$$NH_4OH + HNO_3 \rightarrow NH_4NO_3 + H_2O \qquad (3\text{-}23)$$

The water formed is only slightly ionized, so the solution conductivity is lowered even further.

Chlorides

Another parameter that is carefully monitored and controlled in most nuclear facilities is chloride (Cl^-). The reason for maintaining the chloride ion concentration at the minimum level practicable is that several forms of corrosion are affected by the chloride ion, and the type of greatest concern is chloride stress corrosion. The mechanics of chloride stress corrosion were discussed in detail in Module 2 and will therefore not be repeated in this section.

When high levels of Cl^- are suspected, or detected, immediate steps must be taken to eliminate the source and remove Cl^- from the system because of the potential consequences. If Cl is present in the reactor coolant system, one method of removing it is to initiate a feed and bleed operation after determining that makeup water supplies are not the source of contamination. Because of the large volume of water normally contained in the reactor coolant system, cleanup by this method involves considerable amounts of pure water and a significant amount of time.

Additional problems associated with feed and bleed operations include changes in pH and a loss of H_2 from the reactor coolant system during the cleanup. Changes to either, or both, of these parameters have the potential to further aggravate the occurrence of chloride stress corrosion because: pH changes from the optimum operating limits cause increased general corrosion; and a loss of H_2 from the reactor coolant makes the dissolved oxygen that is normally present from either radiolysis or contained in the makeup water available to interact with the Cl^- ions. This would promote chloride stress corrosion (recalling that Cl^-, O_2, and tensile stress are all necessary for chloride stress corrosion to occur).

The fact that a large amount of makeup water is being introduced will result in hydrogen depletion because even deaerated water contains small amounts of oxygen. H_2 will also be lost because of the draining of coolant from the system. During conditions that require the use of feed and bleed to correct a chemistry anomaly of any type, increased attention to all parameters becomes increasingly important.

Another method that may be used to control Cl^- concentrations in the reactor coolant is to route more bypass flow through the ion exchanger system, where installed. Use of this system precludes the requirements of a large inventory of makeup water as well as the disposal problems that arise from the bleed (draining) of coolant from the system. Because the ion exchanger has minimal effect on dissolved gases, depletion of hydrogen does not contribute to the potential of the problem. Increasing flow through the ion exchanger(s) may cause changes to the pH of the system, however, and as in the case of feed and bleed, increased vigilance is necessary to ensure pH, as well as other parameters, are properly controlled and maintained.

Fluorine

Fluorine (F^-) is another parameter monitored and controlled in some reactor facilities. High levels of F^- are potentially hazardous for two reasons. First, F^- promotes corrosion of zirconium by a stress corrosion mechanism at the inner surface of the clad (fluorine can be introduced to this region because of the existence of small defects or "pinholes" that cannot be completely avoided in the fuel clad). Second, F^- is a major contributor to radiation levels of the reactor coolant. Although some radioactive ^{18}F is produced by a proton, neutron reaction with ^{18}O, the only significant possible sources are impure makeup water or contamination of chemical agents, such as NH_3 and Li, which are added to the reactor coolant system.

$$^{18}O\ (p,n)\ ^{18}F \quad \text{(half-life F 112 minutes)}$$

Removal of F^- is accomplished in the same manner as the removal of Cl, that is, by feed and bleed or increased flow through the ion exchanger system. Removal by ion exchange, however, is not as effective as for Cl^- because of the lower affinity of F^- for exchange. As in the case of Cl^- contamination, if feed and bleed is the method selected for cleanup, the purity of the makeup water must be verified and other chemistry parameters monitored closely to preclude aggravation of the existing conditions.

Radioactivity

Radioactivity of the reactor coolant system is monitored and controlled because it is a good indicator of several conditions that may occur in the system. These include increased corrosion, crud bursts, fuel element defect or failure, and improper operation of the ion exchanger (where installed). Radioactivity levels in the reactor coolant system are normally kept at nominal values during normal conditions by the ion exchanger. In facilities that do not use ion exchangers, other methods that vary from facility to facility are used to control radioactivity. These include processes such as chemical cleanup followed by feed and bleed with subsequent disposal procedures and reactor shutdown in severe cases. Reactor coolant gaseous activity is also monitored, and it provides an indication of cladding failures and the introduction of air during reactor operation. Gaseous activity levels are normally controlled by degassification. Operating limits are established in most reactor facilities to operate the reactor safely, to limit radioactive releases to the environment, and to prevent excessive personnel exposure during operation and shutdown.

Tritium

This section provides introductory information about tritium and its properties.

Environmental Sources

Tritium occurs naturally in the environment. Reactions between cosmic rays and upper atmosphere constituents produce tritium. The following are examples of these reactions.

$${}^{14}_{7}N + {}^{1}_{0}n \rightarrow {}^{3}_{1}H + {}^{12}_{6}C$$

$${}^{2}_{1}H + {}^{2}_{1}H \rightarrow {}^{3}_{1}H + {}^{1}_{1}H$$

Tritium becomes incorporated into water and falls to earth as rain. At an estimated production rate of 4×10^6 Ci/Yr, a world steady state natural inventory of 70×10^6 Ci results.

Man-made Sources

Numerous potential and actual sources of tritium production exist in the United States, the United Kingdom, France, and other countries. They include light-water reactors, heavy-water reactors, fuel reprocessing facilities, and production reactors. Light-water reactors produce between 500 and 1000 Ci/yr of tritium in their coolant for every 1000 MW(e) of power. Heavy-water reactors produce approximately 2×10^6 Ci/yr of tritium in their coolant for every 1,000 MW(e) of power.

Atomic Weight/Hydrogen Isotopes

The atomic weights, symbols, and abundance of the three well-known isotopes of hydrogen are given in Table 2. ${}^4_1H$ and ${}^5_1H$ are also known. However, because they decay in fractions of a single second, they are not extensively studied. Unless otherwise specified in this chapter, the term hydrogen includes protium, deuterium, and tritium. ${}^1_1H$ will be used to refer to protium; confusion with elemental hydrogen will be eliminated by spelling out the latter.

TABLE 2
Hydrogen Isotopes

Physical Symbol	Common Symbol	Name	Abundance (%)	Mass (amu)
${}^1_1H$	H	Protium	99.985	1.007825
${}^2_1H$	D	Deuterium	0.015	2.01400
${}^3_1H$	T	Tritium	β^- emitter*	3.01605

* 12.32-years half-life

Disintegration

Tritium decays by emitting a weak beta particle together with an antineutrino. The product is helium-3. Helium is a monatomic gas; therefore, the decay of 1 mole of T_2 yields 2 moles of helium. This causes a pressure buildup in sealed vessels containing diatomic tritium gas (or HT or DT gas). The following reaction is tritium disintegration.

$$ {}^3_1H \rightarrow \left({}^3_2He\right)^+ + \beta^- + \bar{\nu} $$

Solubility

Hydrogen gas (including tritium gas) dissolves to some extent in most materials. Tritium and deuterium behave like protium, except for small isotopic effects. Hydrogen atoms or diatomic hydrogen molecules and some larger hydrogen-bearing molecules dissolve interstitially; that is, they diffuse into a structure and locate between atoms or molecular frameworks. As it dissolves, the hydrogen slightly disrupts the structural networks of the material and causes expansion. The extent of the disruption, along with the extent of chemical attractive/repulsive forces between the hydrogen and the material, determines the ultimate equilibrium state.

Solubility in Polymers, Glasses, and Ceramics

Hydrogen gas dissolves as the molecular hydrogen species in many materials. For these systems, solubility is generally endothermic and is directly proportional to gaseous overpressure. The materials in which hydrogen dissolves as the molecular species include organic polymers and glasses at less than 150°C. Organic polymers generally have atomic-scale voids in the twisted polymer chain, and the hydrogen molecules seek these sites. Hydrogen dissolution in glasses is assumed to be similar.

Abnormal Chemistry Conditions

As indicated earlier in this module, there are times when actions taken regarding one specific chemistry parameter may affect more than that parameter (such as feed and bleed operations). This may also be the case during many abnormal conditions that the reactor coolant system may be subjected to.

In this section, we will consider the conditions resulting from injection of air, fuel element failure, and overheating of resin and discuss the probable changes to other chemistry parameters that may occur.

Injection of Air

Injection of small and large amounts of air into the reactor coolant system was previously analyzed. When sufficient oxygen is added to deplete the hydrogen inventory, acidic pH results in those facilities where high (basic) pH is maintained. This reduction in pH is accompanied by secondary effects to the coolant. The reduced pH causes a change in solubility of the corrosion film on facility materials and results in part of this film being released for transport throughout the coolant system. This release, or crud burst, is detected by an increase in radiation levels of the coolant, increased conductivity, and lowering of the ion exchanger efficiency. Radiation levels increase because corrosion products that had been attached to core surfaces and were highly activated by the neutron flux are transported to other parts of the system.

Conductivity increases because many of the corrosion products are released in an ionic or radical form and are capable of conducting electricity. In the early stages of the introduction of air, conductivity may exhibit erratic behavior because of the relationship of conductivity and pH. As pH begins to fall, conductivity tends to follow, but the presence of suspended corrosion products offsets the theoretical conductivity. The resultant increase is dependent upon the concentration of corrosion products. Additionally, should pH fall below about 7, conductivity increases as pH is further reduced.

Ion exchanger efficiency decreases because the increased crud inventory consists of both particulate and colloidal products. The ion exchanger is only about 90% effective as a filter, as compared to approximately 100% effectiveness as an ion exchanger, so effluent radioactivity levels increase by a greater proportion. Also, crud in the colloidal form has an even lower probability of being removed by ion exchange or filtration than does particulate crud, and again the effluent radioactivity is increased.

Hydrogen and total gas react as expected. Because air contains primarily N_2 and O_2, the gas inventory in the system increases. Hydrogen levels decrease because of reactions with both O_2 and N_2. Air also contains approximately 1% of $^{40}_{18}Ar$ by volume. When large amounts of air are added to the reactor coolant system, the $^{40}_{18}Ar$ becomes activated by the neutron flux in the core region and becomes radioactive in the form of $^{41}_{18}Ar$, causing an increase in the gaseous activity of the reactor coolant.

Figure 6 illustrates changes to chemistry in an experimental facility that was started up with high oxygen and nitrogen in the system from the deliberate addition of air. As the power level and temperature were increased, nitric acid formed and neutralized the ammonium hydroxide present from previous operation. Conductivity and pH decreased until pH 7 was reached; when the water became acidic the conductivity showed a corresponding increase. With the formation of nitric acid (from the nitrogen present), a corresponding increase in chromate ion was observed (in this case Cr concentrations are directly related to crud inventories and result from chemical attack on facility materials). The process was reversed by adding hydrogen, which combined with the oxygen and reduced the nitrate and chromate ions.

It should be noted, however, that in the above experiment the chromate ions were effectively removed from the coolant by decreasing their solubility. This would not be the case for the majority of the crud. The vast majority of crud would remain suspended in the coolant until it was either removed mechanically (filtered) or settled and redeposited at other locations in the system (including redeposition on the core surfaces).

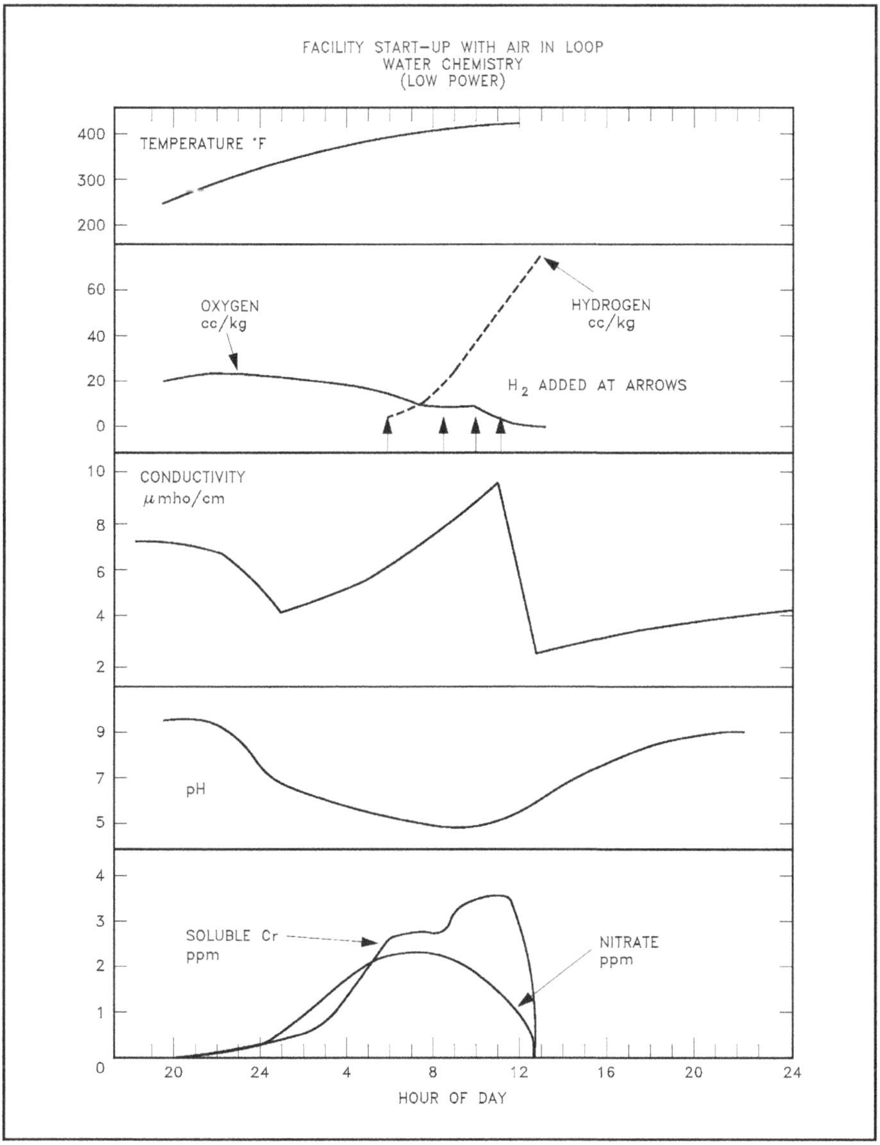

Figure 6 Facility Start-up With Air in Loop

Fuel Element Failure

During operation of a nuclear reactor facility an equilibrium level of fission products is established in the reactor coolant. These fission products are the result of trace impurities of fuel material contained in the cladding surfaces as either natural impurities or a result of the fuel fabrication process. The mechanism by which the fission products enter the coolant is normally by fission recoil. Weld porosity is another potential path for the fuel, but generally quality control prevents this from occurring. During normal facility operation, these fission products are minor contributors to the overall radioactivity of the reactor coolant system.

If a defect were present or a failure of a fuel element occurred, large amounts of fission products would potentially have a path to the coolant system. If this happened, significant changes would occur within the reactor coolant chemistry parameters. Because most facilities analyze for gross coolant radioactivity either continuously or periodically, the analysis would be likely to detect all but the most minute failures.

When routine gaseous radioactive levels are monitored, an increase in this parameter's value would be seen. This is because many of the fission products are gaseous, and these gases are more mobile than particles of exposed fuel (the exposed fuel generally undergoes a process of erosion that washes the fuel into the coolant stream). The other parameter that may change is the ion exchange efficiency (where utilized), because many of the fission products released have a lower affinity for the exchange sites on the resin beads than the exchange anion or cation. Accordingly, the ion exchanger would not effectively remove these fission products and effluent radioactivity levels would increase significantly. Fission gases would also pass on through the ion exchanger and contribute to effluent activity. In addition, because some of the fission gases have relatively short half-lives, the amount of time they are held up in the ion exchanger is sufficient for some of these gases to decay to a radioactive solid.

These solid particles would contribute to effluent samples that may be concentrated prior to analysis. Some facilities monitor for specific fission product inventories in the reactor coolant to provide base level information. If a defect or failure were to occur, these levels would obviously increase to indicate the failure.

Resin Overheating

Because the potential for elevated temperatures exists during most conditions of facility operation, we will summarize the results from the resin in an ion exchanger overheating. Module 4 addresses resin in great detail, and the actual resin breakdown will be included there. Basically the resin of an ion exchanger is an inert polystyrene structure with ion exchange sites "loosely" attached. The basic structure is stable up to fairly high temperatures (approximately 300°F), but the active exchange sites are not. There are two types of exchange sites: anion and cation. The anion resin begins to decompose slowly at about 140°F, and the decomposition becomes rapid above 180°F. The cation resin is stable up to about 250°F. Because these temperatures are well below normal reactor coolant temperatures, the temperature of the coolant must be lowered before it passes through the ion exchange resin.

The decomposition of resin produces an alcohol form of the resin, which has no exchange capability, and trimethylamine (TMA), $N(CH_3)_3$. TMA is a weak base, similar to ammonia, that reacts with water as follows.

$$N(CH_3)_3 + H_2O \leftrightharpoons NH(CH_3)_3^+ + OH^-$$

If large amounts of TMA are released to the coolant, the pH may increase noticeably. For example, 1 ppm of TMA in reactor coolant that uses lithium form resin will cause a noticeable increase in pH. TMA may also interfere with the analysis for chloride ions (which is routinely performed on reactor coolant) by giving a false indication of high chloride concentration. Another significant property of TMA is its intense odor of dead fish. Although the presence of such an odor from reactor coolant is not definitive for TMA, it may give an indication of resin overheating.

The other product of resin breakdown, $[R - CH_2N(CH_3)_2]$, is an amine with exchange capabilities considerably less than the original form of the resin. Thus, both reactions lead to partial (or complete) loss of exchange capability. If the temperature is sufficiently high, or if a lower temperature (but greater than 180°F) is sustained for a long enough period, the resin will be unfit for use. If the temperature becomes very high (greater than about 450°F), the polymeric base structure of the resin will decompose, forming a complex mixture of organic tars and oils.

The preceding discussion concerned the decomposition of resins in their original forms. It should be noted that if overheating occurs after the resin has been in operation for some time, part of the resin will be in a different form due to the exchange process. As a result, some of the previously removed impurities will be released to the coolant if decomposition occurs.

A number of changes are probable if overheating of resin occurs. Reactor coolant Cl^- levels would probably increase as a result of thermal breakdown and subsequent release. Ion exchanger effectiveness would be greatly reduced for similar reasons. Radioactivity levels of the reactor coolant would increase because of the release of impurities collected and later released from the resin. pH would likely decrease because of the release of H^+ ions from the resin complex and may cause acidic conditions in the reactor coolant if the temperature is sufficient (>250°F). Because certain types of resin decompose at lower temperatures, pH may increase as a result of the release of TMA and be accompanied by a dead fish odor. Because of the consequences of overheated resin, stringent temperature limitations are necessary. If overheating occurs, the ion exchanger should be taken out of service immediately and the cause rectified. The resin must be replaced prior to placing the ion exchanger back in service after overheating.

Summary

The important information in this chapter is summarized below.

Chemistry Parameters Summary

- Table 1 provides a summary of the parameters, why they are controlled, and the methods of control.
- The following abnormal chemistry conditions are discussed:

  Injection of air will cause the hydrogen inventory to become depleted, which in turn can lead to a pH swing towards the acidic end. This reduced pH leads to a loosened corrosion film hence a crud burst. The crud burst causes higher radiation levels and an increase in conductivity.

  Fuel element failure allows large amounts of fission products a possible release path to the reactor coolant. This could lead to a lowered ion exchange efficiency.

  Resin overheating results in the decomposition of the resin. The products of the decomposition can seep into the reactor coolant and affect the chemistry in several ways. A few of the common products and their resulting influence on the chemistry is discussed.

Department of Energy
Fundamentals Handbook

CHEMISTRY
Module 4
Principles of Water Treatment

TABLE OF CONTENTS

LIST OF FIGURES

LIST OF TABLES

REFERENCES

- Donald H. Andrews and Richard J. Kokes, Fundamental Chemistry, John Wiley & Sons, Inc., 1963
- Compressed Gas Association, Inc., Handbook of Compressed Gases, 2nd Edition, Reinhold Publishing Corporation, 1981.
- R. A. Day, Jr. and R. C. Johnson, General Chemistry, Prentice Hall, Inc., 1974.
- Dickerson, Gray, Darensbourg and Darensbourg, Chemical Principles, 4th Edition, The Benjamin Cummings Publishing Company, 1984.
- Academic Program for Nuclear Plant Personnel, Volume II, Chemistry, Columbia, MD, General Physics Corporation, Library of Congress Card #A 326517, 1972.
- General Physics Corporation, Fundamentals of Chemistry, General Physics Corporation, 1982.
- Glasstone and Sesonske, Nuclear Reactor Engineering, 3rd Edition, Van Nostrand Reinhold Company, 1981.
- McElroy, Accident Prevention Manual for Industrial Operations Engineering and Technology, Volume 2, 8th Edition, National Safety Council, 1980.
- Sienko and Plane, Chemical Principles and Properties, 2nd Edition, McGraw and Hill, 1974.
- Underwood, Chemistry for Colleges and Schools, 2nd Edition, Edward Arnold, Ltd., 1967.
- Norman V. Steere and Associates, CRC Handbook of Laboratory Safety, 2nd Edition, CRC Press, Inc., 1971.

TERMINAL OBJECTIVE

1.0 Without references, **EXPLAIN** the concept and application of ion exchange in water purification.

ENABLING OBJECTIVES

1.1 **LIST** the three reasons for removing impurities from water prior to use in reactor systems.

1.2 **DEFINE** the following terms:

a. Ion exchange
b. Demineralize
c. Cation
d. Anion
e. Polymer
f. Mixed-bed demineralizer
g. Affinity
h. Decontamination factor

1.3 **DESCRIBE** the following:

a. Resin bead
b. Cation resin
c. Anion resin

1.4 **DISCUSS** the following factors of ion exchange:

a. Relative affinity
b. Decontamination factor

1.5 **WRITE** the reaction for removal of NaCl and $CaSO_4$ by a mixed-bed ion exchanger such as one containing HOH resin.

1.6 **EXPLAIN** the three basic methods used to remove dissolved gases from water.

1.7 **LIST** five filtration mediums used to remove suspended solids from water.

1.8 **EXPLAIN** how mixed-bed ion exchangers may be used to control pH.

ENABLING OBJECTIVES (Cont.)

1.9 **DISCUSS** resin malfunctions, including the following:

a. Channeling
b. Breakthrough
c. Exhaustion

1.10 **LIST** the maximum conductivity and approximate concentration of electrolyte for each level of purity for makeup water.

PURPOSE OF WATER TREATMENT

Water normally contains many impurities, among which are trace minerals and chemicals. The need for removing these impurities is discussed in this chapter.

EO 1.1 LIST the three reasons for removing impurities from water prior to use in reactor systems.

Water Treatment

Water treatment is necessary to remove the impurities that are contained in water as found in nature. Control or elimination of these impurities is necessary to combat corrosion, scale formation, and fouling of heat transfer surfaces throughout the reactor facility and support systems.

The following are three reasons for using very pure water in reactor facility systems.

1. To minimize corrosion, which is enhanced by impurities.

2. To minimize radiation levels in a reactor facility. Some of the natural impurities and most of the corrosion products become highly radioactive after exposure to the neutron flux in the core region. If not removed, these soluble and insoluble substances may be carried to all parts of the system.

3. To minimize fouling of heat transfer surfaces. Corrosion products and other impurities may deposit on core surfaces and other heat transfer regions, which result in decreased heat transfer capabilities by fouling surfaces or blockage of critical flow channels. Areas of high concentrations of these impurities and corrosion products may also lead to extreme conditions of the various corrosion processes with resultant failure of components or systems.

There are several processes used in reactor facilities to purify the water in the systems and water used as makeup. Deaeration is used to strip dissolved gases, filtration is effective in the removal of insoluble solid impurities, and ion exchange removes undesirable ions and replaces them with acceptable ions. Typical ionized impurities found in water are shown in Table 1.

TABLE 1
Typical Ionized Impurities in Water

Cations	Anions
Ca^{++}	NO_3^-
Mg^{++}	OH^-
Na^+	SO_4^{--}
K^+	Cl^-
Al^{+++}	HCO_3^-
Fe^{++}	$HSiO_3^-$
Cu^{++}	$HCrO_3^-$

Summary

The important information in this chapter is summarized below.

Purpose of Water Treatment Summary

- There are three general reasons to treat water for its impurities:

  1. To minimize corrosion, which is enhanced by impurities
  2. To minimize radiation levels in the reactor facility
  3. To minimize fouling of heat transfer surfaces

WATER TREATMENT PROCESSES

One of the more common water treatment methods is the use of demineralizers and ion exchange. This method will be discussed in this chapter.

EO 1.2 DEFINE the following terms:

a.	**Ion exchange**	**e.**	**Polymer**
b.	**Demineralize**	**f.**	**Mixed-bed demineralizer**
c.	**Cation**	**g.**	**Affinity**
d.	**Anion**	**h.**	**Decontamination factor**

EO 1.3 DESCRIBE the following:

a. Resin bead
b. Cation resin
c. Anion resin

EO 1.4 DISCUSS the following factors of ion exchange:

a. Relative affinity
b. Decontamination factor

EO 1.5 WRITE the reaction for removal of NaCl and $CaSO_4$ by a mixed-bed ion exchanger such as one containing HOH resin.

Principles of Ion Exchange

Ion exchange is a process used extensively in nuclear facilities to control the purity and pH of water by removing undesirable ions and replacing them with acceptable ones. Specifically, it is the exchange of ions between a solid substance (called a resin) and an aqueous solution (reactor coolant or makeup water). Depending on the identity of the ions that a resin releases to the water, the process may result in purification of water or in control of the concentration of a particular ion in a solution. An *ion exchange* is the reversible exchange of ions between a liquid and a solid. This process is generally used to remove undesirable ions from a liquid and substitute acceptable ions from the solid (resin).

The devices in which ion exchange occurs are commonly called demineralizers. This name is derived from the term *demineralize*, which means the process whereby impurities present in the incoming fluid (water) are removed by exchanging impure ions with H^+ and OH^- ions, resulting in the formation of pure water. H^+ and OH^- are present on the sites of resin beads contained in the demineralizer tank or column.

There are two general types of ion exchange resins: those that exchange positive ions, called cation resins, and those that exchange negative ions, called anion resins. A *cation* is an ion with a positive charge. Common cations include Ca^{++}, Mg^{++}, Fe^{++}, and H^{+}. A *cation resin* is one that exchanges positive ions. An *anion* is an ion with a negative charge. Common anions include Cl^{-}, SO_4^{-}, and OH^{-}. An *anion resin* is one that exchanges negative ions. Chemically, both types are similar and belong to a group of compounds called *polymers*, which are extremely large molecules that are formed by the combination of many molecules of one or two compounds in a repeating structure that produces long chains.

A *mixed-bed demineralizer* is a vessel, usually with a volume of several cubic feet, that contains the resin. Physically, ion exchange resins are formed in the shape of very small beads, called *resin beads*, with an average diameter of about 0.005 millimeters. Wet resin has the appearance of damp, transparent, amber sand and is insoluble in water, acids, and bases. Retention elements or other suitable devices in the top and bottom have openings smaller than the diameter of the resin beads. The resin itself is a uniform mixture of cation and anion resins in a specific volume ratio depending on their specific gravities. The ratio is normally 2 parts cation resin to 3 parts anion resin.

In some cases, there may be chemical bonds formed between individual chain molecules at various points along the chain. Such polymers are said to be cross-linked. This type of polymer constitutes the basic structure of ion exchange resins. In particular, cross-linked polystyrene is the polymer commonly used in ion exchange resins. However, chemical treatment of polystyrene is required to give it ion exchange capability, and this treatment varies depending on whether the final product is to be an anion resin or a cation resin.

The chemical processes involved in producing anion and cation resins are outlined in Figure 1 and Figure 2, beginning with the formation of cross-linked polystyrene. The polymer itself is a covalent compound. By the chemical reactions indicated in Figure 2, hydrogen atoms covalently bonded to the original polymer at certain sites are replaced by functional groups (called radicals) such as SO_3H (sulfonic acid) and $CH_2N(CH_3)_3Cl$ (quaternary ammonium). Each such group is covalently bonded to the polymer, but each also contains an atom that is bonded to the radical group by a predominantly ionic bond. In the two examples above, H in SO_3H and Cl in $CH_2N(CH_3)_3Cl$ are the ionically-bonded atoms. Sometimes these are written as $SO_3^{-}\ H^{+}$ and $CH_2N(CH_3)_3^{+}Cl^{-}$ to emphasize their ionic characters. These ions ($H^{+}$and Cl^{-}) are replaceable by other ions. That is, H^{+} will exchange with other cations in a solution, and Cl^{-} will exchange with other anions.

In its final form, an ion exchange resin contains a huge, but finite, number of sites occupied by an exchangeable ion. All of the resin, except the exchangeable ion, is inert in the exchange process. Thus, it is customary to use a notation such as R-Cl or H-R for ion exchange resins. R indicates the inert polymeric base structure and the part of the substituted radical that does not participate in exchange reactions. The term R is inexact because it is used to represent the inert portion of both cation and anion resins, which are slightly different. Also, the structure represented by R contains many sites of exchange, although only one is shown by the notation, such as R-Cl. Despite these drawbacks, the term R is used for simplicity.

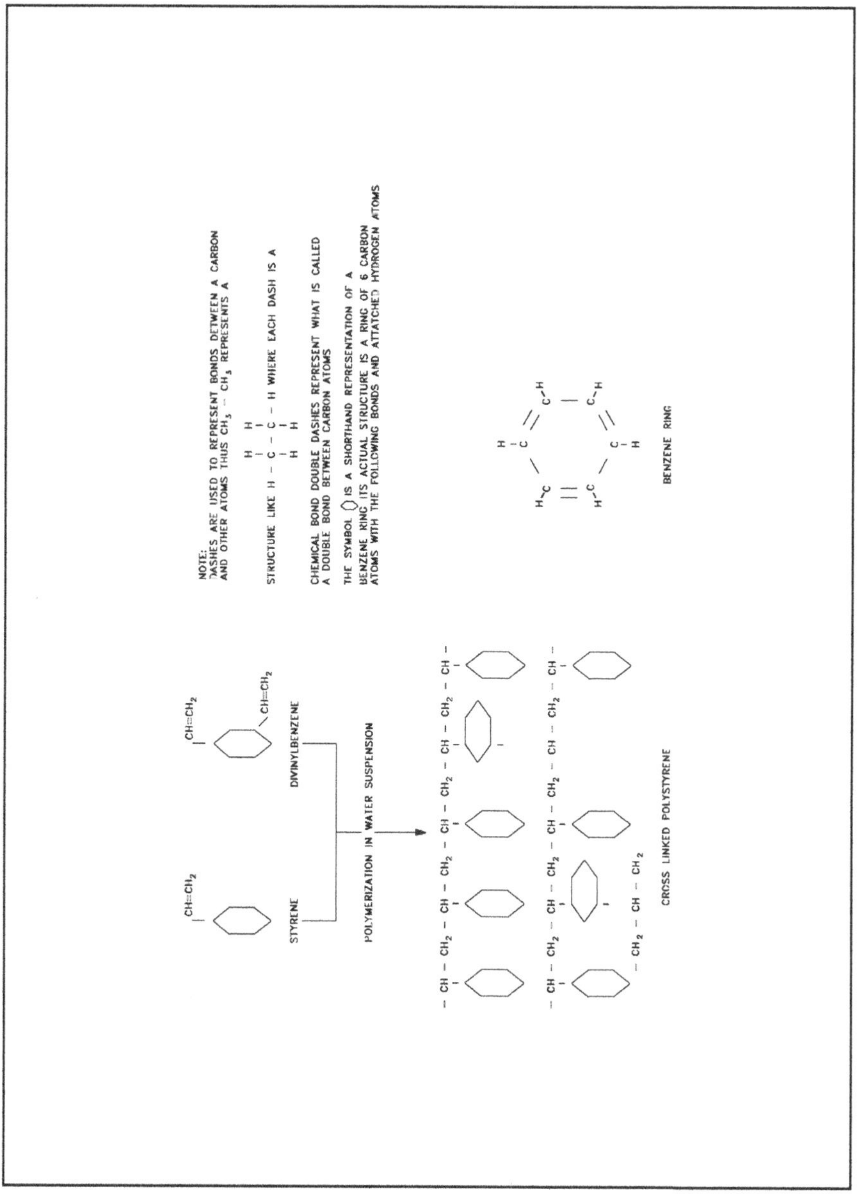

Figure 1 Polymerization of Cross-Linked Polystyrene Resins

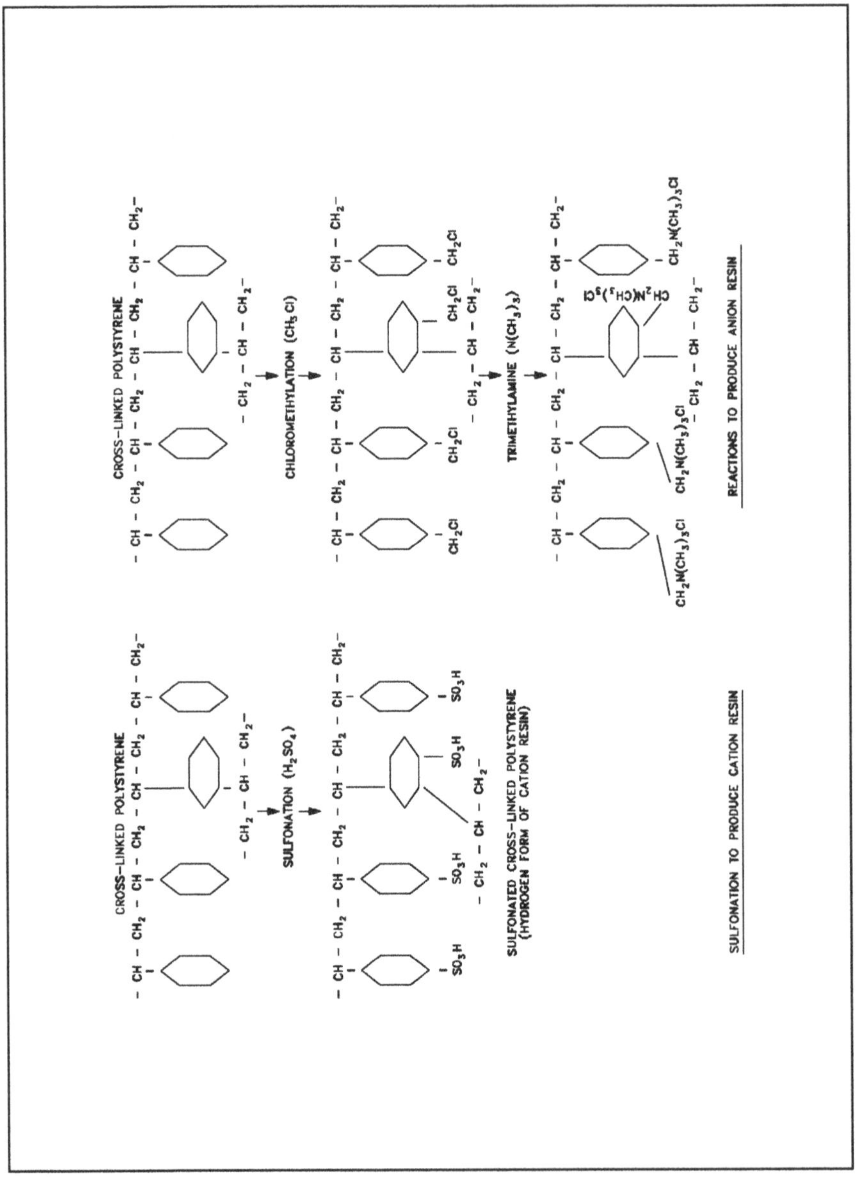

Figure 2 Production of Anion and Cation Forms of Cross-Linked Polystyrene Ion Exchange Resins

A particular resin may be prepared in different forms according to the identity of the exchangeable ion attached. It is usually named according to the ion present on the active sites. For example, the resin represented by R-Cl is said to be the chloride form of the anion resin, or simply the chloride form resin. Other common forms are the ammonium form (NH_4-R), hydroxyl form (R-OH), lithium form (Li-R), and hydrogen form (H-R).

The mechanics of the ion exchange process are somewhat complicated, but the essential features can be understood on the basis of equilibrium concepts discussed in Module 1 and recognition that the strength of the ionic bond between the resin and an ion varies with the particular ion. That is, for a particular resin, different ions experience different attractions to the resin. The term *affinity* is often used to describe the attraction between a resin and a given ion. This affinity can be described quantitatively by experimental determination of a parameter called the relative affinity coefficient. For a qualitative discussion, it suffices to note the *relative affinities* between a resin and different ions.

In order of decreasing strength, the relative affinities between a cation resin and various cations are as follows.

$$Ba^{+2} > Sr^{+2} > Ca^{+2} > Co^{+2} > Ni^{+2} > Cu^{+2} > Mg^{+2} > Be^{+2}$$

$$Ag^{+} > Cs^{+} > Rb^{+} > K^{+} \simeq NH^{+} > Na^{+} > H^{+} > Li^{+}$$

Similarly, the relative affinities between an anion resin and various anions are as follows.

$$SO_4^{-2} > I^{-} > NO_3^{-} > Br^{-} > HSO_3^{-} > Cl^{-} > OH^{-} > HCO_3^{-} > F^{-}$$

The physical arrangement of one type of ion exchange vessel for purifying water is shown in Figure 3. The ion exchange resin is contained in a vessel with a volume of several cubic feet. Retention elements at the top and bottom consist of screens, slotted cylinders, or other suitable devices with openings smaller than the resin beads to prevent the resin from escaping from the vessel. The resin bed is a uniform mixture of cation and anion resins in a volume ratio of 2 parts cation resin to 3 parts anion resin. This arrangement is called a mixed-bed resin, as opposed to an arrangement of cation and anion resins in discrete layers or separate vessels. The use of different volumes of the two types of resins is due to the difference in exchange capacity between cation and anion resins. *Exchange capacity* is the amount of impurity that a given amount of resin is capable of removing, and it has units of moles/ml, equivalents/ml, or moles/gm. The anion resin is less dense than the cation resin; thus, it has a smaller exchange capacity, and a larger volume is needed for anion resins than for the cation resins to obtain equal total exchange capabilities.

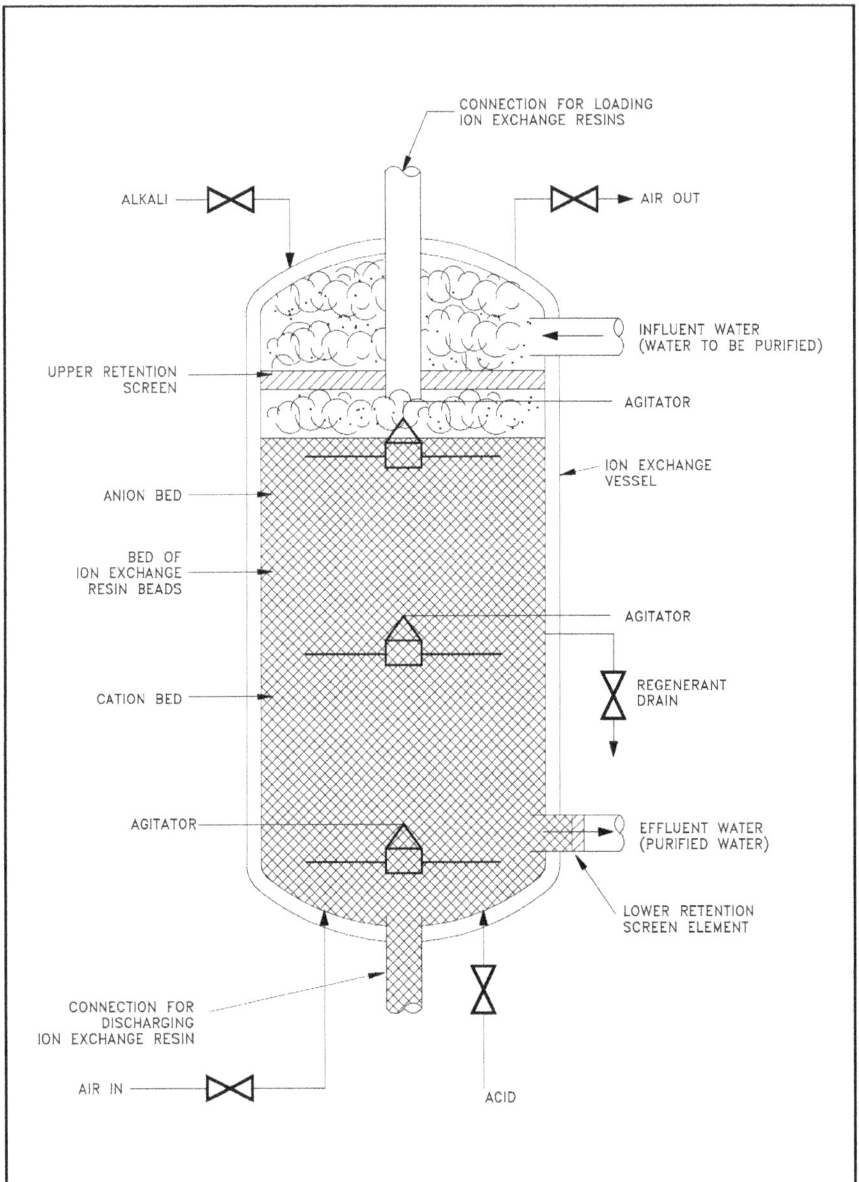

Figure 3 Schematic Diagram of a Typical Ion Exchanger

Because of the different densities of anion and cation resins, the flow of solution (impure water) is from top to bottom. If the flow were reversed, the lighter anion resin would gradually rise to the top by a process called classification, resulting in a layer of anion resin on top of the cation resin, as shown in Figure 3. In the example shown, the layering results from regeneration and/or backwash. In systems not using a backwash, the anion and cation resin beads are uniformly mixed. Many systems use a backwash procedure, if the resins are regenerated, to remove solids collected by filtration and to separate the resins for regeneration. They are remixed after regeneration.

For fixed amounts of anion and cation resins, the efficiency for removal of impurities is greater in a mixed-bed resin than a layered arrangement. The main reason is that for layered resins there may be large pH gradients within the column of resin. If, for example, the hydroxyl form resin is on top, as solution passes through it anionic impurities are removed and replaced by OH^- ions; thus, the pH increases. This increase in pH may decrease the efficiency in lower portions of the resin bed for removing impurities. It may also cause some impurities to precipitate because solubility changes with pH. The resin column will filter some undissolved material, but the efficiency for filtration is usually significantly less than that for removal by ion exchange. Thus, the overall efficiency is less than in a mixed-bed resin.

The capacity of ion exchange resins to remove impurity ions is given in Table 2 along with other information on resins. For instance, each cubic foot of a mixed-bed resin is capable of exchanging with 19.8 moles each of monovalent cations and anions. Mixed-bed resins are available commercially and in practical applications several cubic feet are used in a purification system.

TABLE 2
Properties of Ion Exchange Resins

Properties	Cation Resin	Anion Resin	Mixed-Bed Resin
Ion exchange capacity, moles of single ion/ml	1.75×10^{-3}	1.20×10^{-3}	0.7×10^{-3} anion & cation
Ion exchange capacity, moles of single ion/ft^3	49.5	34.0	19.8 anion & cation
Density of wet resin particles, grams/ml	1.27	1.10	--
Bulk density of loaded bed (including voids), grams/ml	0.80	0.62	0.70
Volume fraction	--	--	60% anion & 40% cation

The ion exchange process is reversible. If too much solution is passed through the ion exchanger (that is, the capacity of the resin has been exceeded) the exchange may reverse, and undesirable ions or other substances that were previously removed, will be returned to the solution at the effluent. Therefore, it is necessary to periodically monitor the performance of the ion exchanger and either replace or regenerate the resin when indicated. Ion exchanger performance is measured by comparing the solution concentration, conductivity, or radioactivity at the influent and effluent. The parameter measured depends upon the purpose of the ion exchanger. The term normally applied to ion exchanger effectiveness is *decontamination factor* (DF), which is defined as a ratio of the concentration (or activity) of the fluid at the inlet compared to the concentration (or activity) at the effluent, which expresses the effectiveness of an ion exchange process.

$$DF = \frac{\text{Influent concentration, conductivity, or radioactivity}}{\text{Effluent concentration, conductivity, or radioactivity}}$$

Example 1:

An ion exchanger influent contains 15 ppm chloride (Cl). Effluent chloride is measured at 0.1 ppm. What is the DF of this ion exchanger?

Solution:

$$DF = \frac{\text{Influent concentration}}{\text{Effluent concentration}}$$

$$DF = \frac{15 \text{ ppm Cl}}{0.1 \text{ ppm Cl}}$$

$$DF = 150$$

Example 2:

Reactor coolant activity entering the purification ion exchanger equals 2.8×10^{-2} µCi/ml gross activity. Ion exchanger effluent activity taken at the same time and conditions is measured at 1.0×10^{-4} µCi/ml. What is the DF?

Solution:

$$DF = \frac{\text{Influent radioactivity}}{\text{Effluent radioactivity}}$$

$$DF = \frac{2.8 \times 10^{-2} \ \mu\text{Ci/ml}}{1.0 \times 10^{4} \ \mu\text{Ci/ml}}$$

$$DF = 280$$

Resin performance may be monitored using a history curve that plots DF with respect to time. A typical history curve is shown in Figure 4, with the resin considered "exhausted" at point X.

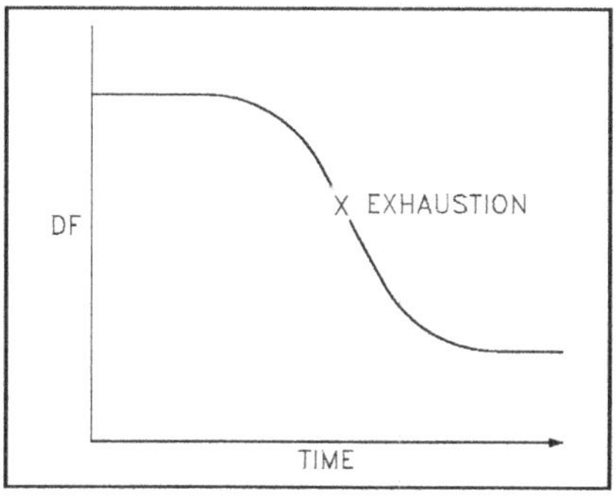

Figure 4 Typical History Curve

Specific Ion Exchanger Reactions

Suppose a solution containing Na^+ ions is passed through hydrogen resin. From the relative affinities given earlier, Na^+ ions are attracted to the resin more strongly than H^+ ions. Thus, Na^+ ions will displace H^+ ions from the resin or, in other words, Na^+ ions and H^+ ions exchange place between resin and solution. The process can be described by the following equilibrium reaction.

$$H{-}R + Na^+ \rightleftharpoons Na{-}R + H^+ \qquad (4\text{-}1)$$

In most practical situations, a solution containing impurities at low concentrations is passed through a large amount of resin. By LeChatelier's Principle, the equilibrium of Reaction (4-1) is forced far to the right. The equilibrium is displaced so far that, for practical purposes, all Na^+ ions are removed from solution and replaced by H^+ ions. As a result, the solution will be acidic because of the excess of H^+ ions.

If a solution containing Cl^- ions is passed through hydroxyl resin, the Cl^- ions will be removed according to the following reaction.

$$R{-}OH + Cl \rightleftharpoons R{-}Cl + OH \qquad (4\text{-}2)$$

Again, for a dilute solution and a large amount of resin, the removal of Cl^- ions is essentially 100 percent complete. In this case, the final solution will be basic because of the excess of OH^- ions.

Consider a situation in which the entering impurities are calcium sulfate ($CaSO_4$) and sodium chloride (NaCl), and the ion exchanger is a mixture of both hydrogen and hydroxyl resins (mixed-bed).

$$H{-}R + R{-}OH + Ca^{++} + SO_4 \rightarrow Ca{-}R + R{-}SO_4 + H^{+} + OH^{-} \quad (4\text{-}3)$$

$$H{-}R + R{-}OH + Na^{+} + Cl^{-} \rightarrow Na{-}R + R{-}Cl + H^{+} + OH^{-} \quad (4\text{-}4)$$

In the reaction with NaCl, both Na^+ and Cl^- ions are removed from solution and replaced by H^+ and OH^- ions, respectively (the $CaSO_4$ reaction has the same result). If the initial solution contained only NaCl, then the concentrations of Na^+ and Cl^- ions were equal. Because both are removed with 100 percent efficiency, the concentrations of H^+ and OH^- ions added to the solution are equal; thus, the solution is neutral. In solution, H^+ and OH^- ions must obey the relationship for the ionization of water (refer to Module 1).

$$K_w = 10^{-14} = [H^+] [OH^-]$$

Because of the very small value of the dissociation constant K_w, the great majority of H^+ and OH^- ions supplied by the resin must combine to form water by the following reaction.

$$H^+ + OH^- \rightarrow H_2O \quad (4\text{-}5)$$

By this process, the original NaCl solution becomes pure water.

The preceding examples involve hydrogen and hydroxyl resins. The use of other resins, especially cation resins, is very common. For instance, suppose a solution containing Na^+ ions is passed through a lithium resin. Again referring to the relative affinities, Na^+ is attracted to the resin more strongly than is Li^+; thus, Na^+ ions will displace Li^+ from the resin.

Summary

The important information in this chapter is summarized below.

Ion Exchange Summary

- Demineralize is defined as the process whereby impurities present in the incoming fluid (water) are removed by exchanging impure ions with H^+ and OH^- ions resulting in the formation of pure water.

- Ion exchange is a process used extensively in nuclear facilities to control the purity and pH of water by removing undesirable ions and replacing them with acceptable ones.

- Mixed-bed demineralizer is a vessel containing resin that is a uniform mixture of cation and anion resins in a specific volume ratio depending on their specific gravities. Normally the ratio is 2 parts cation resin to 3 parts anion resin.

- Cation is an ion with a positive charge. Common cations include Ca^{++}, Mg^{++}, Fe^{++}, and H^+. A cation resin is one that exchanges positive ions.

- Anion is an ion with a negative charge. Common anions include Cl^-, SO_4^{-2}, and OH^-. An anion resin is one that exchanges negative ions.

- Decontamination factor (DF) is a ratio of the concentration (or activity) of the fluid at the inlet compared to the concentration (or activity) at the effluent, which defines the effectiveness of the ion exchange process. Mathematically it is:

$$DF = \frac{\text{Influent concentration, conductivity, or radioactivity}}{\text{Effluent concentration, conductivity, or radioactivity}}$$

- Polymers are extremely large molecules that are formed by the combination of many molecules of one or two compounds in a repeating structure that produces long chains.

- Affinity is often used to describe the attraction between a resin and a given ion. This affinity can be described quantitatively by experimental determination of a parameter called the relative affinity coefficient.

Ion Exchange Summary (Cont.)

- Resin beads are long-chain cross-linked polymers that contain sites occupied by exchangeable ions. The general order of affinity serves as a useful guide in understanding many ion exchange processes. Cation and anion resins are named according to the identity of the ion occupying the exchange sites, such as hydrogen, hydroxyl, and chloride. Mixed-bed resins are used to remove both cations and anions.

- The effectiveness of any ion exchanger is directly related to the relative affinities between a resin and different ions. In order of decreasing strength, the relative affinities between a cation resin and various cations are:

  $Ba^{+2} > Sr^{+2} > Ca^{+2} > Co^{+2} > Ni^{+2} > Cu^{+2} > Mg^{+2} > Be^{+2}$

  $Ag^{+} > Cs^{+} > Rb^{+} > K^{+} \simeq NH^{+} > Na^{+} > H^{+} > Li^{+}$

  Similarly, the relative affinities of the anion resin for various anions are:

  $SO_4^{-2} > I^{-} > NO_3^{-} > Br^{-} > HSO_3^{-} > Cl^{-} > OH^{-} > HCO_3^{-} > F^{-}$

  The higher the relative affinity the more effective the ion exchanger. This effectiveness is expressed by the Decontamination Factor.

- The reaction for removal of NaCl and $CaSO_4$ by a mixed-bed ion exchanger such as one containing HOH resin is as follows:

$$H{-}R + R{-}OH + Ca^{++} + SO_4^{=} \rightarrow Ca{-}R + R{-}SO_4 + H^{+} + OH$$

$$H{-}R + R{-}OH + Na^{+} + Cl \rightarrow Na{-}R + R{-}Cl + H^{+} + OH$$

DISSOLVED GASES, SUSPENDED SOLIDS, AND pH CONTROL

The presence of dissolved gases, suspended solids, and incorrect pH can be detrimental to the water systems associated with a reactor facility. Therefore, these conditions must be minimized or eliminated to reduce corrosion in the systems of the facility. The way these conditions are controlled and the difficulties in controlling them are discussed in this chapter.

EO 1.6 **EXPLAIN the three basic methods used to remove dissolved gases from water.**

EO 1.7 **LIST five filtration mediums used to remove suspended solids from water.**

EO 1.8 **EXPLAIN how mixed-bed ion exchangers may be used to control pH.**

EO 1.9 **DISCUSS resin malfunctions, including the following:**

- **a. Channeling**
- **b. Breakthrough**
- **c. Exhaustion**

Removal of Dissolved Gases

Dissolved gases result from different sources depending upon which system we examine. In the following discussion, we will address makeup water, reactor coolant systems, secondary facility water systems, the sources of dissolved gases, and methods used to reduce their concentrations to acceptable levels.

Many facilities use raw water as a source for makeup water systems. Pretreatment of this water is accomplished in various ways from distillation to a series of distinct processes as shown in Figure 5. In a pretreatment system similar to that shown in Figure 5, a resin column containing a cation resin (hydrogen form) is used to remove cations. The water entering the cation exchanger contains numerous ions including sodium (Na^+), bicarbonate (HCO_3^-), and others (HCO_3^- is one of the major impurities in many raw water systems). Na^+ ions result from the water softener located upstream in the pretreatment system. In addition to the HCO_3^- ions, raw water contains large amounts of magnesium (Mg^{++}) and calcium (Ca^{++}), as well as small amounts of other ionic impurities.

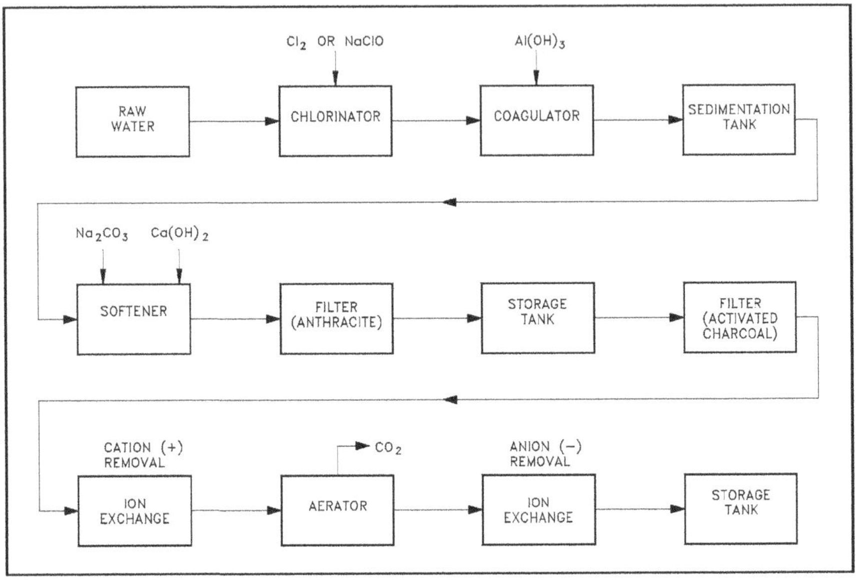

Figure 6 A Typical Pretreatment System

The reactions that occur in the water softener include the removal of both Mg^{++} and Ca^{++} ions. The water softener contains resin in which the insoluble exchange site is the SO_3^- molecule, and the soluble ions attached to the exchange site are Na^+ ions. When water containing Mg^{++}, Ca^{++}, and HCO_3^- ions is passed over the resin in the softener, the ions are exchanged by the following reaction (Mg^{++} removal is similar).

$$\underset{\text{(resin complex)}}{2R{-}SO_3Na^+} + Ca^{++} + HCO_3^- \rightarrow \underset{\text{(resin complex)}}{2R{-}SO_3Ca^{++}} + 2Na^+ + HCO_3^- \qquad (4\text{-}6)$$

Note that electrical neutrality is maintained before and after the exchange reaction. One calcium ion with two positive charges is attached to two exchange sites that release two sodium ions with one positive charge each. The HCO_3^- ion is not affected by the reaction and passes through the resin of the softener.

To obtain pure water, it is necessary to demineralize the water completely, which is accomplished using a cation exchanger, an aerator, and an anion exchanger.

The cation exchanger contains resin in the hydrogen form. In this treatment step, essentially all cations entering the ion exchanger will be held at the exchange site, and H^+ will be released as shown in the following typical reaction (the anions, specifically the HCO_3^- ions, are unaffected by the cation exchanger).

$$Na^+ + HCO_3^- + R{-}SO_3^-H^+ \rightarrow R{-}SO_3^-Na^+ + H^+ + HCO_3^- \qquad (4\text{-}7)$$

The water leaving the resin is somewhat acidic (depending on the incoming ion concentration) because it contains H^+ ions and whatever anion was associated with the incoming cation. After passing through the cation resin, the HCO_3^- ions combine with the H^+ ions to form carbonic acid (H_2CO_3). Carbonic acid is a weak acid that will decompose to water and CO_2 by the following reaction.

$$H_2CO_3 \leftrightarrows H_2O + CO_2 \qquad (4\text{-}8)$$

Because the carbonic acid readily dissociates, the aerator is used to remove the CO_2 from the makeup water at this point in the system. If we aerate the water by some means, such as spraying it through a tower or blowing air through the water, the CO_2 is "stripped" from the water and vented to the atmosphere. The removal of CO_2 forces Reaction (4-8) to shift to the right, which converts more H_2CO_3 to CO_2. With sufficient aeration, all bicarbonate (HCO_3^-), and therefore CO_2, can be removed.

Similar reactions occur in the anion exchanger. For example, anion resin, which has hydroxide ions at the exchange sites, will react as indicated in the following typical reaction.

$$H^+ + Cl^- + R{-}N(CH_3)_3^+OH^- \rightarrow R{-}N(CH_3)_3^+Cl^- + HOH \qquad (4\text{-}9)$$

In this pretreatment system, the anion resin is downstream of the cation resin, and the only cations present are hydrogen ions. When the hydroxyl ions are released from the anion exchange sites, they combine with the hydrogen ions to form water. As a result, pure water appears at the effluent (this is somewhat overstated because a very small amount of other cations and anions pass unaffected through the resin columns in actual practice).

Another method sometimes used to remove dissolved gases from water is deaeration. In this process, the water is stored in vented tanks containing electric heaters or steam coils. The water is heated to a temperature sufficient for slow boiling to occur. This boiling strips dissolved gases from the stored water, and these gases are then vented to the atmosphere. Usually, the vented gases are directed through a small condenser to limit the loss of water vapor that would escape as steam along with the gases. This method is particularly effective in removing dissolved oxygen as well as other entrained gases (CO_2, N_2, and Ar).

Removal of dissolved gases from the reactor coolant system is usually accomplished by venting a steam space or high point in the system. In pressurized water reactors (PWR), this is normally accomplished in the pressurizer. The steam space is the high point of the system, and the boiling and condensing action causes a constant stripping of dissolved gases to occur. The steam space is vented either intermittently or constantly, and the gases are carried off in the process.

In addition to the mechanical means mentioned above, the use of scavengers in a PWR prevents the presence of dissolved oxygen. Two methods are normally used in this regard. When facility temperature is above approximately 200°F, gaseous hydrogen is added and maintained in the primary coolant to scavenge oxygen by the following reaction.

$$2H_2 + O_2 \rightarrow 2H_2O \qquad (4\text{-}10)$$

The other scavenger is hydrazine (N_2H_4). Hydrazine is thermally unstable and decomposes at temperatures above 200°F to form ammonia (NH_3), nitrogen (N_2), and hydrogen (H_2). Consequently, the use of hydrazine as an oxygen scavenger is limited to temperatures below 200°F. Hydrazine scavenges oxygen by the following reaction.

$$N_2H_4 + O_2 \rightarrow 2H_2O + N_2\uparrow \qquad (4\text{-}11)$$

The presence of dissolved gases in the steam facility of a PWR is as detrimental as the presence of these gases is in the reactor coolant systems. Because steam facility systems contain metals other than stainless steel, they are even more susceptible to certain types of corrosion in the presence of oxygen and carbon dioxide. Removal of dissolved gases from the steam system is accomplished in two ways: by mechanical means such as air ejectors or mechanical pumps; and by using chemicals that scavenge oxygen.

Because boiling occurs in the steam generators, any dissolved gases entrained in the feedwater will be stripped out during the boiling process. These gases are carried with the steam through the turbines and auxiliary systems and ultimately end up in the condensers. The design of the condensers is such that noncondensible gases (for example, O_2, CO_2) are collected and routed to the air removal system (which consists of air ejectors or mechanical pumps), where they are subsequently discharged to the atmosphere.

Scavenging involves the use of solid additives and volatile chemicals. One commonly-used solid chemical additive is sodium sulfite (Na_2SO_3). Scavenging of oxygen occurs by the following reaction.

$$2Na_2SO_3 + O_2 \rightarrow 2Na_2SO_4 \qquad (4\text{-}12)$$

As can be seen by Reaction (4-12), oxygen is consumed in the reaction resulting in the formation of sodium sulfate, Na_2SO_4 (a soft sludge). Addition of this scavenging agent is limited to drum-type steam generators. Once Thru Steam Generators (OTSG) do not use this method, but instead use controls that keep all scale-forming chemicals out of the steam generators.

Sodium sulfite reacts rapidly with oxygen and is a very efficient scavenger. However, being a solid and the source of another solid (Na_2SO_4) that is produced during the reaction, sodium sulfite has the potential of fouling heat transfer surfaces. An additional problem associated with the use of sodium sulfite is corrosion of secondary system components resulting from its decomposition products. At the temperatures present in the steam generators, sodium sulfite can decompose as follows.

$$H_2O + Na_2SO_3 \rightarrow 2NaOH + SO_2 \quad (4\text{-}13)$$

Sulfur dioxide (SO_2) is a gas and is carried over to the remainder of the steam facility. With water (in the steam or in the feed/condensate system), the SO_2 reacts in the following manner.

$$H_2O + SO_2 \rightarrow \underset{\text{(acidic)}}{H_2SO_3} \quad (4\text{-}14)$$

This acidic condition is corrosive to all components in the secondary system.

Because of the problems associated with sodium sulfites, many facilities use volatile chemistry control of the secondary steam system to control dissolved gases in conjunction with air removal systems. This control utilizes hydrazine (Reaction 4-11) and morpholine (Reaction 4-15) to eliminate oxygen and carbon dioxide, respectively.

$$N_2H_4 + O_2 \rightarrow H_2O + N_2\uparrow \quad \text{(oxygen consumed)} \quad (4\text{-}11)$$

$$2C_4H_9NO + CO_2 + 2H_2O \rightarrow C_4H_9NO \bullet C_4H_9COOH + HNO_3 + H_2 \quad \text{(carbon dioxide consumed)} \quad (4\text{-}15)$$

As can be seen by Reaction (4-11), no solids are formed; thus, the tendency of fouling heat transfer surfaces is reduced. An additional benefit results from the decomposition of hydrazine by the following reactions.

$$2N_2H_4 \rightarrow 2NH_3 + N_2 + H_2 \quad (4\text{-}16)$$

$$NH_3 + H_2O \rightarrow NH_4OH \quad (4\text{-}17)$$

These reactions result in an alkaline pH condition that decreases corrosion in the steam facility. As can be seen in Reaction (4-15), the consumption of CO_2 takes place. Two benefits result from this reaction; 1) the inventory of dissolved gases in the steam facility is reduced, and 2) is the reaction contributes to maintaining a higher pH by eliminating carbonic acid (H_2CO_3), thus reducing corrosion.

Removal of Suspended Solids

Referring back to Figure 5 and examining the effluent of the softener, we find that both sodium salts and precipitates are present. These substances result from reactions that typically occur based on the presence of Ca^{++} and Mg^{++} salts. The chemicals most commonly used for softening are soda ash or sodium carbonate (Na_2CO_3) and hydrated lime ($Ca(OH)_2$). Hard water (water containing Ca^{++} and Mg^{++} salts) contains calcium and magnesium bicarbonates ($Ca(HCO_3)_2$) and ($Mg(HCO_3)_2$), as well as calcium sulfate ($CaSO_4$) and magnesium chloride ($MgCl_2$). These impurities produce the following reactions.

$$Ca(HCO_3)_2 + Ca(OH)_2 \rightarrow 2CaCO_3 + 2H_2O \quad (4\text{-}18)$$

$$Mg(HCO_3)_2 + 2Ca(OH)_2 \rightarrow Mg(OH)_2 + 2CaCO_3 + 2H_2O \quad (4\text{-}19)$$

$$MgSO_4 + Ca(OH)_2 \rightarrow Mg(OH)_2 + CaSO_4 \quad (4\text{-}20)$$

$$CaSO_4 + Na_2CO_3 \rightarrow CaCO_3 + Na_2SO_4 \quad (4\text{-}21)$$

$$MgCl_2 + Ca(OH)_2 \rightarrow Mg(OH)_2 + CaCl_2 \quad (4\text{-}22)$$

$$CaCl_2 + Na_2CO_3 \rightarrow CaCO_3 + 2NaCl \quad (4\text{-}23)$$

As evident from the above reactions, although Ca^{++} and Mg^{++} ions can be removed from solution, soluble sodium salts are formed. Consequently the total dissolved solid content is essentially unchanged. $CaCO_3$ and $Mg(OH)_2$ are in precipitate form and must also be removed from the solution. One method of doing this is by filtration. Filtration is the process in which insoluble solids are removed from the water by passing them through a filter medium consisting of some type of porous material. This process will remove suspended solids and precipitates, but has no effect on dissolved solids. Numerous materials are used as filter media and include sand, activated charcoal, anthracite, diatomaceous earth, and to some extent resin in an ion exchanger. Sand is not normally used in nuclear applications because of the silicate ion ($SiO_3^{=}$) associated. Silicate ions are undesirable because they hydrolyze in water and form a weak acid, which tends to increase corrosion. Activated charcoal is often used following a chlorinator in a water treatment system because it removes excess residual chlorine as well as suspended matter.

There are two types of mechanical filters in use, gravity and pressure. Pressure filters are the most widely used because they can be installed in a pressurized system, thereby eliminating the need for additional pumps (gravity filters require pumps to provide a motive force). In addition, in a pressurized filter system, flow rate and other associated parameters can be better controlled.

Another method used for the removal of suspended corrosion products in facility fluid systems is the electromagnetic filter. These are gaining popularity in PWR feed and condensate systems where they have proven effective in reducing the crud loading of these systems, thereby reducing the inventory of corrosion products in steam generators.

Ion exchangers also function as filtration units by virtue of the size of the resin beads and the torturous path the water must follow in passing through the resin. The filtration efficiency, however, is significantly less than the ion exchange efficiency (90% or less for filtration versus approximately 100% for most ion exchange reactions). Filtration efficiency depends largely on the size of the suspended materials, with greater efficiency for the larger particles. Adverse effects of this filtration process are similar to those occurring in other types of filters. In a radioactive system, the buildup of filtered particles (crud) can increase the radiation to prohibitive levels or cause flow reductions that may necessitate removal or backwash of the resin.

pH Control

As discussed in Module 2, in reactor facilities other than those containing aluminum components or using chemical shim, reactor coolant is maintained in an alkaline condition to control corrosion in the system. In practice, if the desired alkaline condition is established, and no other action is taken, the pH gradually decreases during operation of the facility. This is due to factors such as dilution from makeup water additions to compensate for coolant losses caused by sampling, leakage, volume changes on facility cooldown, and reaction of hydroxyl ions (OH^-) with metals and corrosion products. Thus, hydroxyl ions must be added to the coolant to maintain a basic pH. The ion exchange process provides a convenient means of helping to control the pH of reactor coolant by adding OH^- ions from the exchange reactions of the resin.

A hydroxyl form anion resin, and a cation resin in some form other than the hydrogen form, may be used for this purpose as in a mixed-bed arrangement, similar to that shown in Figure 3, or in separate cation and anion units where flow is individually controlled. Usually, a portion of the reactor coolant is diverted from its normal path, passes through the ion exchange resin, and then is returned to the main coolant path. In this way, part of the coolant is constantly purified, and in the process hydroxyl ions are released to the coolant. These hydroxyl ions tend to increase the pH (or limit the decrease), thereby offsetting the effects mentioned in the preceding paragraph.

Suppose that the cation resin is in the ammonium form. As reactor coolant passes through the ion exchange system, ionic impurities will be removed and NH_4^+ and OH^- will be added. Thus, the resin serves both to purify the coolant and to help maintain the pH of the coolant by releasing NH_4OH.

The major sources of ionic impurities in reactor coolant are impurities in makeup water and corrosion products. The amount of these impurities in reactor coolant is normally very small; thus, the base added to the coolant by the ion exchange resin is usually not sufficient to entirely compensate for the losses described above. For this reason, the pH of reactor coolant is measured regularly, and additional base is added as needed. The frequency of addition varies considerably with the type of base used.

Resin Bed Malfunctions

The resin beds of ion exchangers are susceptible to malfunction from a number of causes. These causes include channeling, breakthrough, exhaustion, and overheating.

Channeling

Channeling is a condition in which the resin allows a direct flow of water through the ion exchanger. Flow channels are established from the inlet to the outlet of the ion exchanger, which allows water to flow essentially unrestricted through the resin via these paths. If channeling occurs, the water flowing through the resin bed has insufficient contact with the resin beads and results in a decrease in effectiveness of the ion exchanger.

Channeling most often results from improper filling of the ion exchanger with resin. If insufficient water is mixed with the resin when it is added, the resin column may contain pockets, or voids. These voids may then set up flow paths for channeling to occur. Improper design or malfunction of the water inlet connection (flow diffuser) can also lead to channeling.

Breakthrough and Exhaustion

To gain further insight into the processes that occur in a column of mixed bed resin as it removes an impurity, it is worthwhile to construct a series of curves such as those shown in Figure 6. These curves illustrate the behavior of hydrogen form resin as it removes NH_4^+ ions from solution (the behavior of an anion resin is analogous). The two rows of curves schematically represent the concentrations of H^+ and NH_4^+ on the resin and in the effluent (exiting solution).

In the bottom row, concentration is plotted against volume of solution that has passed through the column. That is, the plots represent the concentration of the indicated ion in the effluent solution after a volume of solution has passed through the resin. It is assumed that the concentration of NH_4^+ in the influent is constant. NH_4^+ ions are more strongly attracted to the resin than are H^+ ions. Thus, NH_4^+ ions readily exchange for H^+ ions on the resin.

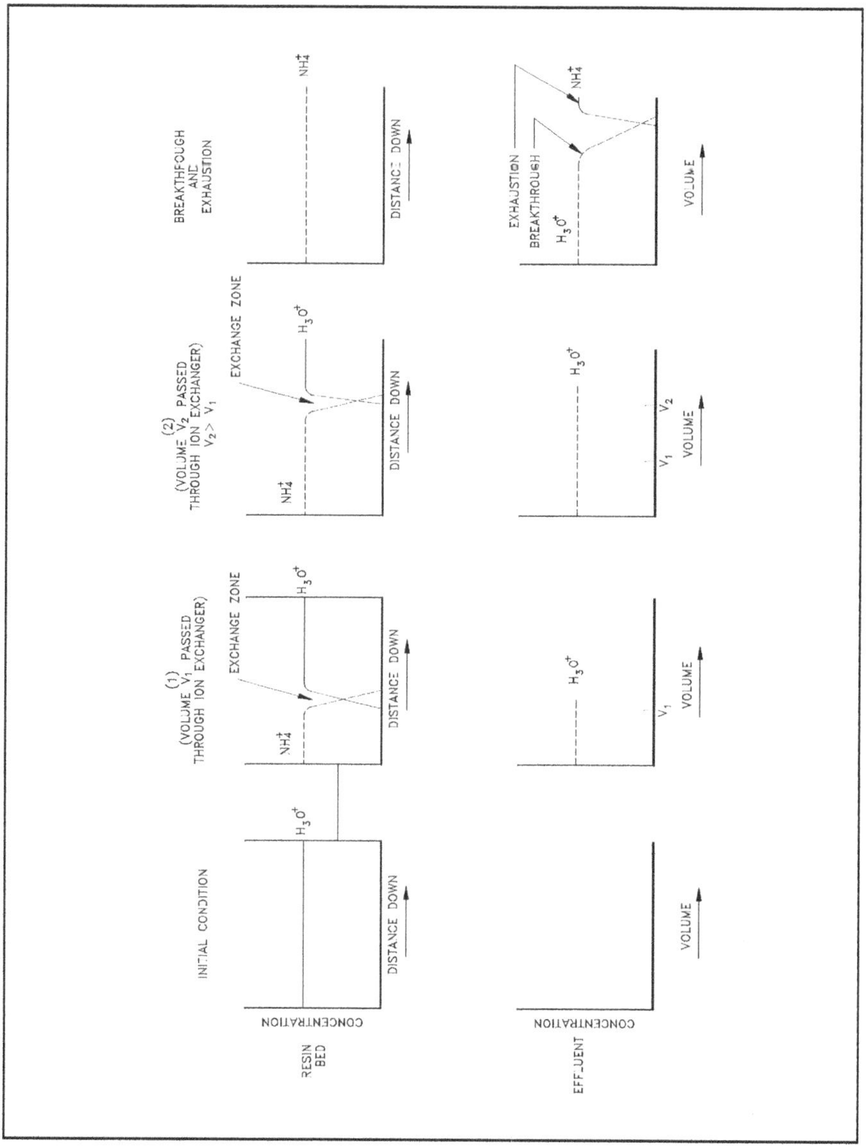

Figure 6 Behavior of Hydrogen Form Cation Exchange Bed as it Removes Ammonium Ions From Solution

As solution passes through the resin column, the relative amounts of ammonium and hydrogen ions on the resin change. The actual exchange process occurs primarily in a relatively narrow band of the column rather than over the entire length. This band is called the exchange zone. Assuming the column is vertical and that solution flows from top to bottom, the resin above the exchange zone is depleted; that is, practically all the exchange capacity has been used. Below the exchange zone, essentially none of the resin's exchange capacity has been used. As more and more solution flows through the column, the exchange zone gradually moves downward as more of the resin is depleted. Eventually, as the exchange zone approaches the end of the column, small amounts of NH_4^+ begin to appear in the effluent. The point at which this occurs is called breakthrough. If more solution passes through the resin, the concentration of NH_4^+ in the effluent increases until it is the same as the concentration in the influent. This condition is called exhaustion and indicates that essentially all the exchange capacity of the resin has been used. (Because of the equilibrium nature of the exchange process, a small amount of the resin may remain in the hydrogen form, but not enough to remove any more ionic impurities.) Note that because the exchange zone in this case was narrow, a relatively small volume of solution takes the resin from breakthrough to exhaustion.

Resin Overheating

The potential for elevated temperatures exists during most conditions of facility operation, we will examine in detail the processes that occur if the resin in an ion exchanger is overheated. Although the inert polystyrene basic structure of resin is stable up to fairly high temperatures (approximately 300°F), the active exchange sites are not. The anion resin begins to decompose slowly at about 140°F, and the decomposition becomes rapid above 180°F. The cation resin is stable up to about 250°F. Because these temperatures are well below normal reactor coolant temperatures, the temperature of the coolant must be lowered before it passes through the ion exchange resin.

The anion resin (hydroxyl form) decomposes by either of two mechanisms with approximately equal probability.

$$R-CH_2N(CH_3)_3^+OH^- \xrightarrow[\text{heat}]{} R-CH_2OH + N(CH_3)_3 \quad (4\text{-}24)$$

or

$$R-CH_2N(CH_3)_3^+OH^- \xrightarrow[\text{heat}]{} R-CH_2N(CH_3)_2 + CH_3OH \quad (4\text{-}25)$$

Reaction (4-24) produces an alcohol form of the resin, which has no exchange capability, and trimethylamine (TMA), $N(CH_3)_3$. TMA is a weak base, similar to ammonia, that reacts with water as follows.

$$N(CH_3)_3 + H_2O \rightleftharpoons NH(CH_3)_3^+ + OH^-$$

If large amounts of TMA are released to the coolant, the pH may increase noticeably. For example, 1 ppm of TMA in reactor coolant that uses lithium resin will cause a noticeable increase in pH. TMA may also interfere with the analysis for chloride ions (which is routinely performed on reactor coolant) by giving a false indication of high chloride concentration. Another significant property of TMA is its intense odor of dead fish. Although the presence of such an odor from reactor coolant is not definitive for TMA, it may give an indication of resin overheating.

The methyl alcohol (CH_3OH) produced by Reaction (4-25) is not expected to have a harmful effect on the reactor coolant system. The other product of this reaction [R - $CH_2N(CH_3)_2$] is an amine with exchange capabilities considerably less than the original form of the resin. Thus, both reactions lead to partial (or complete) loss of exchange capability. If the temperature is sufficiently high, or if a lower temperature (greater than 180°F) is sustained for a long enough period, the resin will be unfit for use.

Cation exchange resin begins to undergo thermal decomposition at temperatures above about 250°F by the following reaction.

$$R{-}SO_3H + H_2O \xrightarrow[heat]{} R{-}H + 2H^+ + SO_4^= \quad (4\text{-}26)$$

This reaction destroys all exchange capacity of the cation resin and also produces an acid. The Reactions (4-24) through (4-26) are the initial reactions when resin is overheated. If the temperature becomes very high (greater than about 450°F), the polymeric base structure of the resin will decompose, forming a complex mixture of organic tars and oils.

The preceding discussion concerned the decomposition of resins in their original forms. It should be noted that if overheating occurs after the resin has been in operation for some time, part of the resin will be in a different form due to the exchange process. As a result, some of the previously-removed impurities will be released to the coolant if decomposition occurs. For example, the chloride form of the anion resin will form CH_3Cl by the reaction corresponding to Reaction (4-25).

$$R{-}CH_2N(CH_3)_3^+Cl^- \xrightarrow[heat]{} R{-}CH_2N(CH_3)_2 + CH_3Cl$$

The CH_3Cl (chloromethane) will be released to the coolant and will decompose in the radiation field of the reactor core, producing Cl^- ions. Similarly, the sodium form of the cation resin will release Na^+ ions by the following reactions.

$$R-SO_3Na + H_2O \xrightarrow[heat]{} R-H + H^+ + Na^+ + SO_4^=$$

A number of changes are probable if overheating of resin occurs. Reactor coolant Cl^- levels would probably increase as a result of thermal breakdown and subsequent release. Ion exchanger effectiveness would be greatly reduced for similar reasons. Radioactivity levels of the reactor coolant would increase because of the release of impurities collected and later released from the resin. pH would likely decrease because of the release of H^+ ions from the cation resin complex and may cause acidic conditions of the reactor coolant if the temperature were sufficient (>250°F). Because the anion resin decomposes at lower temperatures, pH may increase as a result of the release of TMA and be accompanied by a dead fish odor. Because of the consequences of overheated resin, stringent temperature limitations are necessary. If overheating occurs, the ion exchanger should be taken out of service immediately and the cause rectified. The resin must be replaced prior to placing the ion exchanger back in service after overheating.

Summary

The important information in this chapter is summarized below.

Dissolved Gases, Suspended Solids And pH Control Summary

- Because of the presence of impurities, raw water sources undergo treatment prior to use as makeup water in reactor facility systems. These systems normally utilize several distinct processes that remove solids, ionic impurities, and gases. Pretreatment of makeup water is necessary to reduce corrosion, minimize radiation, and limit fouling of heat transfer surfaces.

- If the hydrogen and hydroxyl forms of resin are used, the result is pure, neutral water. Other forms may be used to remove unwanted impurities from solution and substitute another substance, such as a base, to help control pH. Resins also filter solids suspended in a solution, but the efficiency for this is usually less than the efficiency for removal by ion exchange.

Dissolved Gases, Suspended Solids And pH Control Summary (Cont.)

- Dissolved gases are removed from reactor facility systems to limit corrosion by any one of the following methods or combinations of methods:

  Aeration - The spraying of the water to physically release the entrained gasses, then venting the gas.

  Deaeration - The heating of the water to a slow boil, then vents the gas usually to a condenser so the moisture is not lost.

  Addition of scavengers - Examples are hydrogen, hydrazine, and morpholine, the scavenger combines with the gas and removes it chemically.

- Channeling in an ion exchanger occurs when there is a direct flowpath for the water that decreases the resin-water contact. Channeling may occur in the resin bed as a result of improper filling or malfunction of the inlet connection.

- Resins are susceptible to damage by overheating. Although a resin contains a very large number of exchange sites, the number is finite and the resin exchange capability is eventually depleted. The first indication of depletion is breakthrough, which occurs when impurity ions begin to appear in the solution after it has passed through the resin.

- When the concentration of impurities is the same before and after the solution passes through the resin, the resin has reached a state of exhaustion.

- Suspended solids are removed by the use of mechanical filters. The two basic categories of mechanical filters are gravity flow and pressure flow filters. The pressure flow filters are more likely to be used because there are better control capabilities. Numerous materials are used as filter media and include sand, activated charcoal, anthracite, diatomaceous earth, and to some extent the resin in an ion exchanger

WATER PURITY

The fewer the contaminants in water, the less corrosion takes place. The methods of water treatment have been explored previously. This chapter discusses how water purity is quantified. Measuring the purity helps to keep treatment effective.

EO 1.10 LIST the maximum conductivity and approximate concentration of electrolyte for each level of purity for makeup water.

Water Purity

The water used in a nuclear facility must be of a purity level that is consistent with the overall objectives of chemistry control in the facility.

There are a number of ways in which pure water is obtained, including distillation systems and pretreatment systems similar to those mentioned earlier in this module. Regardless of the method employed, the required purity must be achieved.

Water purity has been defined in many different ways, but one generally accepted definition states that high purity water is water that has been distilled and/or de-ionized so that it will have a specific resistance of 500,000 ohms (2.0 micromhos conductivity) or greater. This definition is satisfactory as a base to work from, but for more critical requirements, the breakdown shown in Table 3 has been suggested to express degrees of purity.

TABLE 3
Water Purity

Degree of Purity	Maximum Conductivity (µmhos/cm)	Approximate Concentration of Electrolyte, mg/1
Pure	10	2 - 5
Very Pure	1	0.2 - 0.5
Ultrapure	0.1	0.01 - 0.02
Theoretically Pure	0.054	0.00

Conductivity is a measure of the ease with which electricity can be passed through a substance. The presence of ions greatly facilitates the passage of an electric current. Pure water is only slightly ionized by the dissociation of water: $H_2O \rightarrow H^+ + OH^-$. At 25°C, the concentration of the hydrogen and hydroxyl ions is 10^{-7} moles/liter.

The equivalent conductance of hydrogen (H) is

$$350 \ \frac{\text{mhos-cm}^2}{\text{equivalent}}$$

and the equivalent conductance of OH is

$$192 \ \frac{\text{mhos-cm}^2}{\text{equivalent.}}$$

A mho is a measure of the ease with which electric current will pass and is the inverse of an ohm, the measure of resistance to the passage of electric current. Conductance and conductivity are similar qualities (conductivity is measured in μmhos/cm, so conductance must be converted to conductivity). A μmho is one millionth of a mho. The total conductivity of pure water can be calculated by adding the equivalent conductances of H and OH, multiplying by the normality (see Module 1 for definition), and then multiplying by 10^{3} l/cm^3 and 10^6 μmhos/mho. For theoretically pure water this becomes the following.

$$(350 + 192) \frac{\text{mhos} - \text{cm}^2}{\text{equiv}} \times 10^{-7} \frac{\text{equiv}}{\text{liter}} \times 10^{-3} \frac{\text{liters}}{\text{cm}^3} \times 10^6 \frac{\mu\text{mhos}}{\text{mho}} = .054 \frac{\mu\text{mho}}{\text{cm}}$$

The conductivity limit for demineralized water is 1 μmho/cm.

Conductivity will very quickly indicate the presence of any ionic impurities, even if the impurity concentration is extremely small. As an example, suppose 1.0 mg of NaCl impurity were deposited in 1 liter of demineralized water. The normality of this solution would be as follows.

$$1 \times 10^{3} \frac{\text{gm}}{\text{liter}} \times \frac{1 \text{ equivalent}}{58 \text{ gm}} = \frac{1}{58} \times 10^{3} \frac{\text{equivalent}}{\text{liter}} = 1.7 \times 10^{5} \frac{\text{equivalent}}{\text{liter}}$$

The equivalent conductance of Na^+ is $51 \ \frac{\text{mhos} - \text{cm}^2}{\text{equivalent}}$, while the equivalent conductance of Cl^- is $75 \ \frac{\text{mhos} - \text{cm}^2}{\text{equivalent}}$. The conductivity of the solution is

$$(51 + 75)\ \frac{\text{mhos} - \text{cm}^2}{\text{equivalent}} \times 1.7 \times 10^{-5}\,\frac{\text{equivalent}}{\text{liter}} \times 10^{-3}\,\frac{\text{liter}}{\text{cm}} \times 10^{6}\,\frac{\mu\text{mho}}{\text{mho}} = 2.2\,\frac{\mu\text{mho}}{\text{cm}},$$

which is well above the limit. Even extremely low concentrations of ionic impurities can easily be detected. The most probable cause is a depleted or damaged resin bed that is no longer capable of removing ionic contaminants. The depleted bed should be removed from service, and a fresh resin bed placed in service.

For most applications in nuclear facilities, the specification that is identified as Very Pure (1.0 μmho/cm maximum conductivity) is used. Ultra Pure demineralized water is normally only required in laboratory situations and is mentioned for information purposes only.

Summary

The important information in this chapter is summarized below.

Water Purity Summary

Water Purity is clarified below.

Degree of Purity	Maximum Conductivity(μmhos/cm)	Approximate Concentration of Electrolyte, mg/1
Pure	10	2 - 5
Very Pure	1	0.2 - 0.5
Ultrapure	0.1	0.01 - 0.02
Theoretically Pure	0.054	0.00

Department of Energy
Fundamentals Handbook

CHEMISTRY
Module 5
Hazards of Chemicals and Gases

TABLE OF CONTENTS

LIST OF FIGURES

LIST OF TABLES

REFERENCES

- Donald H. Andrews and Richard J. Kokes, Fundamental Chemistry, John Wiley & Sons, Inc., 1963
- Compressed Gas Association, Inc., Handbook of Compressed Gases, 2nd Edition, Reinhold Publishing Corporation, 1981.
- R. A. Day, Jr. and R. C. Johnson, General Chemistry, Prentice Hall, Inc., 1974.
- Dickerson, Gray, Darensbourg and Darensbourg, Chemical Principles, 4th Edition, The Benjamin Cummings Publishing Company, 1984.
- Academic Program for Nuclear Plant Personnel, Volume II, Chemistry, Columbia, MD, General Physics Corporation, Library of Congress Card #A 326517, 1972.
- General Physics Corporation, Fundamentals of Chemistry, General Physics Corporation, 1982.
- Glasstone and Sesonske, Nuclear Reactor Engineering, 3rd Edition, Van Nostrand Reinhold Company, 1981.
- McElroy, Accident Prevention Manual for Industrial Operations Engineering and Technology, Volume 2, 8th Edition, National Safety Council, 1980.
- Sienko and Plane, Chemical Principles and Properties, 2nd Edition, McGraw and Hill, 1974.
- Underwood, Chemistry for Colleges and Schools, 2nd Edition, Edward Arnold, Ltd., 1967.
- Norman V. Steere and Associates, CRC Handbook of Laboratory Safety, 2nd Edition, CRC Press, Inc., 1971.

TERMINAL OBJECTIVE

1.0 Without references, **DISCUSS** the hazards associated with chemicals (liquid and gas) found in a nuclear plant.

ENABLING OBJECTIVES

1.1 **STATE** the hazards associated with the use of corrosives.

1.2 **STATE** the general safety precautions necessary for the handling, storage, and disposal of corrosives.

1.3 **LIST** the general safety precautions regarding toxic compounds.

1.4 **LIST** the criteria used to determine if a compound is a health hazard.

1.5 **STATE** the methods by which toxic compounds may enter the body.

1.6 **SUMMARIZE** the purpose and general contents of the following:

a. Material Safety Data Sheets (MSDS)
b. Toxic Substance List

1.7 **DEFINE** the following terms:

a. Compressed gas
b. Non-liquified gases
c. Liquified gases
d. Dissolved gases

1.8 **STATE** the five major families of gases.

1.9 **STATE** the general safety precautions regarding the use, handling, and storage of gases.

1.10 **STATE** the safety precautions for working with cryogenic liquids.

1.11 **LIST** the physical properties and special precautions for the following gases:

a. Hydrogen
b. Oxygen
c. Nitrogen

ENABLING OBJECTIVES (Cont.)

1.12 **DEFINE** the following terms:

a. Flammable liquid
b. Combustible liquid

1.13 **STATE** general safety precautions regarding the use, handling, and storage of flammable and combustible liquids.

1.14 **STATE** the reasons for and techniques used in bonding and grounding of flammable liquid containers.

1.15 **LIST** four sources of ignition of flammable liquids.

1.16 **STATE** the health hazards associated with flammable and/or combustible liquids.

CORROSIVES (ACIDS AND ALKALIES)

There are two basic groups of corrosives: acids and alkalies. These chemicals require precautions for safe handling. These precautions will be discussed in this chapter.

EO 1.1 STATE the hazards associated with the use of corrosives.

EO 1.2 STATE the general safety precautions necessary for the handling, storage, and disposal of corrosives.

Acids

Acids are compounds of hydrogen and one or more other elements (with the exception of carbon) that dissociate or break down to produce hydrogen ions (H^+) when dissolved in water or certain other solvents.

Acids are corrosive in any form, and in high concentrations destroy body tissue and cause severe burns on contact with the skin. The eyes are very susceptible, and permanent damage or loss of sight may result from contact with acids. The inhalation of excessive concentrations of vapor or mist is extremely irritating to the respiratory system and to mucous membranes in particular. Accidental swallowing of concentrated acids may result in severe irritation of, and damage to, the throat and stomach which, in some cases, may prove fatal. Some of these materials are specifically poisonous as well as irritating. In lower concentrations, repeated skin contact may result in inflammation.

Concentrated aqueous solutions of acids are not in themselves flammable. The potential hazard is the danger of their mixture with other chemicals or combustible materials which may result in fire or explosion. Acids also react with many metals resulting in the liberation of hydrogen, a highly flammable gas, which upon ignition in air may cause an explosion. Some of the acids are strong oxidizing agents and can react destructively and violently when in contact with organic or other oxidizable materials.

Personnel exposure requiring immediate action usually involves direct contact of the acid with the body or eyes of the individual, inhalation of acid vapors or decomposition products, and ingestion of acid. The initial treatment in all cases of local contact is immediate removal of the acid with a large amount of water. This treatment must be prolonged until all traces of acid have been removed, usually a minimum washing time of 15 minutes.

Alkalies

Alkalies (bases) are corrosive caustic substances that dissociate in water and yield hydroxyl ions (OH^-). Alkalies include: ammonia, ammonium hydroxide; calcium hydroxide and oxide; potassium, potassium hydroxide and carbonate; sodium, sodium hydroxide; carbonate, peroxide and silicate; and trisodium phosphate.

The alkalies, whether in solid form or concentrated liquid solution, are more destructive to tissue than most acids. Alkali dusts, mists, and sprays may cause irritation of the eyes and respiratory tract and lesions of the nasal septum. Strong alkalies combine with tissue, causing severe burns, frequently deep ulceration, and ultimate scarring. Severe burns result not only from contact with solid alkalies, but also from solutions of these compounds. Potassium and sodium hydroxide are the most active materials in this group. Even dilute solutions of the stronger alkalies tend to soften the epidermis (skin) and emulsify or dissolve the skin fats. Exposure to atmospheres contaminated with alkalies may result in damage to the upper respiratory tract and to lung tissue, depending upon the severity of the exposure. The effects of inhalation may vary from mild irritation of the nasal mucous membranes to severe inflammation of the lungs.

Ingestion causes severe damage to mucous membranes or deeper tissues with which contact is made. Perforation of these tissues may follow, or there may be severe and extensive scar formation. Death may result if penetration into vital areas occurs.

Even though alkalies are not flammable and will not support combustion, much heat is evolved when the solid material is dissolved in water. Therefore, cold water must be used to dissolve solid alkalies, otherwise the solution may boil, and splatter corrosive liquid over a wide area.

General Safety Precautions

Corrosives are available in numerous forms and varying concentrations. Some forms and concentrations are more hazardous than others, but the potential for serious accidents exists regardless of the substance in question.

Many of the safety precautions necessary for safe handling and storage are equally applicable to acids and alkalies. Some of the more common precautions are contained in this section. These precautions are not all inclusive, nor are they meant to be. Specific corrosives may require specific precautions, and Material Safety Data Sheets (MSDS) must be consulted in all cases. The MSDS will be discussed later in this module.

Safety in handling hazardous chemicals depends to a great extent upon effective employee education, proper safety practices, intelligent supervision, and the use of safe equipment. Workers should be thoroughly informed of the hazards that may result from improper handling. Each employee should know what to do in an emergency and should be fully informed about proper first-aid measures. Hazards from spills and leaks should be minimized by an adequate supply of water for washing-down. Drainage of hard-surfaced or diked areas should be directed to minimize the exposure of personnel and equipment. Adequate ventilation should be provided in areas where chemical mist or dust is present.

Alkalies are much more injurious to the eyes than acids because strong acids tend to precipitate a protein barrier, which prevents further penetration into the tissue. The alkalies do not do this. They continue to soak into the tissue as long as they are allowed to remain in contact with the eye. The end result of a corrosive burn to the eye (alkali or acid) is usually a scar on the cornea and possible permanent damage.

Speed in removing corrosives is of primary importance. If the chemical enters the eyes, they should be copiously irrigated with water for at least 15 minutes, and a physician should be consulted immediately. In case of contact with skin or mucous membranes, the safety shower should be used immediately. Clothing can be removed under the shower. Contaminated skin areas should be washed with very large quantities of water for 1 to 2 hours, or until medical help arrives. The ready availability of water, particularly safety showers and eye-washing baths, greatly minimizes the possibility of severe, extensive damage. Contaminated clothing and shoes should be thoroughly washed and decontaminated before re-use.

The use of personal protective equipment is not intended as a substitute for adequate control measures, but because corrosives can cause extensive damage to the body this equipment must be available as needed. During handling operations where spills or splashes are possible, whole body protection (eyes, head, body, hands, and feet) may be necessary. All personal protective equipment should be carefully cleaned and stored following use, and any equipment that cannot be decontaminated should be discarded.

For the protection of the eyes, chemical safety goggles should be worn. Face shields should be worn if complete face protection is necessary. Eyewash fountains and safety showers must be available at any location where eye and/or skin contact may occur. Protection against mist or dust can be provided by proper respiratory protective equipment. The wearing of protective clothing is also advisable to avoid skin contact. This may consist of rubber gloves, aprons, shoes or boots, and cotton coveralls which fit snugly. Safety shoes or boots made of rubber, chlorobutadiene, or other chemical-resistant materials with built-in steel toecaps are recommended for workers handling drums or in process areas where leakage may occur. Containers should be stored in rooms with trapped floor drains. Curbs or a drained gutter, covered with an appropriate grill, should be constructed at door openings where floor drains are not provided.

Tanks should be entered for cleaning or repairing only after these have been drained, flushed thoroughly with water, ventilated, and sampled. Workers entering tanks should be monitored by someone on the outside of the tank. A supplied-air respirator or self-contained breathing apparatus, together with rescue harness and lifeline, should be on hand for rescue purposes.

Removal from exposure is the primary, and most important, step where exposure by inhalation is involved. The individual should be made as warm and comfortable as possible, and a physician should be called immediately.

Ingestion, the least common mode of contamination, requires immediate medical attention. Any attempt at first aid beyond drinking large quantities of water should be made only upon the advice of a physician.

If body burns are severe or extensive, or if the eyes are in any way involved, a physician should be consulted as soon as possible after first aid is rendered. No attempt should be made to neutralize the corrosive prior to treatment with water. Any treatment, in addition to the use of water, should be undertaken only with the advice of the physician.

When corrosives are shipped in small containers such as glass or polyethylene bottles, they should be well protected, whether individually packaged or several are packaged in a single case. After careful inspection, the corrosives may be stored in these containers if the containers are maintained in an upright position and under cover. The containers should be kept off the floor on materials that are corrosive resistant, or protected with corrosive-resistant coverings, to facilitate flushing and other cleanup procedures in the event of leakage or spills.

All drums should be stored on individual racks or securely blocked on skids, with the closure (plug) up to prevent leakage. Drums containing corrosives in liquid form should be vented when received, and at least weekly thereafter, to relieve accumulated internal pressure.

Cylinders should be stored in an upright position, preferably in individual racks and with the valve protective cap in place. In all cases, to avoid error, empty and full containers should be stored in different locations.

Under no circumstance should corrosives be transferred from the original labeled container to an unmarked container. All containers must be labeled clearly, concisely, and in simple, easily understood terms. Inspection of containers before handling will disclose conditions such as breakage, leakage, and improperly positioned closures which could readily cause a leak or spill.

In handling bottles, barrels, or drums containing corrosives, the following guidelines must be followed.

1. Carefully inspect containers prior to handling.
2. Use personal protective equipment.
3. Use equipment specifically designed for the purpose of transporting and dispensing the chemical in question.
4. Label all containers into which the chemical is transferred.

Properties of corrosives make several considerations mandatory in the selection of a storage site.

1. The building, or area within the building selected, should be of fire-resistant construction.

2. The floors should be composed of chemical-resistant brick or treated concrete, be washable, and be provided with adequate drainage.

3. A well-lit and ventilated area in which there are adequate outlets for water should be provided.

4. A relatively cool and dry environment should be maintained, preventing extremes of temperature and humidity.

5. Electrical fixtures should be protected against corrosive mists, and wiring should be enclosed and covered with corrosive-resistant material.

The nature of the corrosive will determine the manner in which it is stored. Most acids should, to some extent, be isolated, some from all other chemicals, some from certain other acids and oxidizable materials such as oil and grease, and some from combustible materials.

Generally, adequate natural ventilation is sufficient in areas where corrosives are stored, that is, where the containers remain unopened. Where acid is used in work areas where dust or mists may arise (such as in processing equipment or tanks), some form of mechanical exhaust system must be provided.

Transporting containers within the plant and dispensing at various points throughout the plant are two high-risk procedures that may cause an accident. Proper equipment can be readily obtained, which precludes the necessity of using makeshift or otherwise dangerous methods of transportation.

Handtrucks or power trucks used for transporting containers should have lifting parts, or clamps specially designed for that purpose. If bottles must be transported in the plant or laboratory, they should be enclosed in safety bottle carriers that prevent breakage if the bottle is struck or dropped. All containers (especially acid) must be opened slowly and cautiously because of the possible buildup of pressure within the container. Corrosives may be dispensed from drums by means of siphons, drum transfer pumps, or by gravity with the use of a special fitting, such as a self-closing faucet. Under no circumstances should bottles or drums be subjected to air pressure to expel the contents.

One final, and extremely important, consideration is the type of container or receptacle into which corrosives are to be dispensed. The use of an inappropriate or makeshift receptacle can negate the value of all precautionary measures.

These receptacles may be used for temporary storage or merely as a means of transporting from storage area to place of use. In any event, an appropriate receptacle meets several conditions.

1. It is designed for the application.
2. It is used for no other purpose than that for which it is intended.
3. It is maintained in a safe, clean, and dry condition.

Summary

The important information in this chapter is summarized below.

Corrosives Summary

- The hazards of acids:

    High concentrations can destroy body tissue, eyes being especially susceptible with permanent damage or loss of sight.

    Inhalation of acidic vapors can irritate the respiratory system.

    Ingestion can destroy the stomach and throat lining, and if the concentration is strong enough, ingestion can be fatal.

    Aqueous solutions can become explosive if combined with other chemicals or combustible materials

    If reacting with metal, hydrogen gas may be a byproduct, which is very explosive.

Corrosives Summary (Cont.)

- The hazards of alkalies:

  Alkalies are more destructive than the acids.

  Alkali dusts, mists, and sprays can cause irritation of nasal passages, eyes, and respiratory tract.

  When in contact with the tissue, strong alkalies will cause ulcers, severe burns, and eventual scarring.

  Ingestion causes perforations of the mucous membrane and deeper tissues; death may result if penetration is in vital areas.

- Precautions when using corrosives:

  An adequate supply of washdown water must be available.

  Proper ventilation in corrosive work area must be provided.

  Proper drainage must be provided such that exposure is limited.

  Face shields and safety glasses that protect the eyes from splashes and extensive vapor should be worn.

  Proper personnel safety equipment should be worn when appropriate (chemical gloves, respirators, coveralls, etc.)

- Precautions when storing corrosives:

  The building, or area within the building selected, should be of fire-resistant construction.

  The floors should be composed of chemical-resistant brick or treated concrete, be washable, and be provided with adequate drainage.

  A well-lit and ventilated area in which there are adequate outlets for water should be provided.

  A relatively cool and dry environment should be maintained, preventing extremes of temperature and humidity.

  Electrical fixtures should be protected against corrosive mists, and wiring should be enclosed and covered with corrosive-resistant material.

TOXIC COMPOUNDS

The various chemicals found in industry as well as the home are useful when properly applied. If the user is uninformed about correct applications, storage, and potential hazards, these chemicals become threats to safety. This chapter gives an overview of handling chemicals and resources of information about these chemicals.

EO 1.3 **LIST the general safety precautions regarding toxic compounds.**

EO 1.4 **LIST the criteria used to determine if a compound is a health hazard.**

EO 1.5 **STATE the methods by which toxic compounds may enter the body.**

EO 1.6 **SUMMARIZE the purpose and general contents of the following:**

a. Material Safety Data Sheets (MSDS)
b. Toxic Substance List

Toxic Compounds

Because the types of toxic compounds found in industry number in the thousands, and because specific hazards, controls, and corrective measures may vary with the substance, no attempt will be made in this section to address specific compounds. Instead, information of a general nature will be presented on toxic materials. Material Safety Data Sheets (MSDS) are required for all potentially hazardous and toxic materials and should be consulted for specific descriptions and precautions concerning the substance in question.

There are some general precautions that should be universally employed regarding toxic compounds. Many of these precautions are consistent with those already mentioned concerning corrosives. Proper ventilation, appropriate hygienic practices, housekeeping, protective clothing, and training for safe handling and storage will diminish many of the hazards that exist.

The toxicity of a material is not synonymous with its health hazard. *Toxicity* is the capacity of a material to produce injury or harm to a living organism.

Hazard is the possibility that a material will cause injury when a specific quantity is used under specific conditions. Several key elements are considered when evaluating a health hazard.

- Toxicity of the materials used
- Physical properties of these materials
- Absorption probabilities of these materials by individuals
- Extent and intensity of exposure to these materials
- Control measures used

Toxicity is relative. It refers to a harmful effect on some biologic mechanism. The term toxicity is commonly used in comparing one chemical agent with another, but such comparison is meaningless if the biologic mechanism, and the conditions under which the harmful effects occur, are not specified.

Although the toxic effects of many chemical agents used in industry are well known, the toxic effects of many other commonly used chemical agents are not as well defined. The toxicity of a material is not a physical constant (such as boiling point, melting point, or temperature); therefore, only a general statement can be made concerning the harmful nature of a given chemical agent.

Many chemical agents are nonselective in their action on tissue or cells; they may exert a harmful effect on all living matter. Other chemical agents may act only on specific cells. Another agent may be harmful only to certain species; other species may have built-in protective devices.

The degree to which a substance will affect living cells can be measured only after recognizable changes have occurred following absorption. Some changes (impaired judgment, delayed reaction time) may be produced at levels too low to cause actual cell damage. Toxicity is dependent upon the dose, rate, method, and site of absorption, and many other factors including general state of health, individual differences, tolerance, diet, and temperature.

In general, industrial poisonings usually result from inhalation, ingestion, and absorption.

- The inhalation and absorption of toxic agents by the lungs is dependent upon the solubility in body fluids, the diffusion through the lungs, the volume of inhalation, the volume of blood in the lungs, and the concentration gradient of vapors between the inhaled air and the blood.
- Ingestion of the toxic agent can occur to some extent; however, there would generally be considerable inhalation of the material where such conditions exist.

- Absorption through the skin can occur upon exposure to some toxic agents. Some liquids and vapors are known to pass through the skin in concentrations high enough such that respiratory protection is not adequate. For example, hydrogen cyanide (HCN) is known to pass through the unbroken skin. Consideration should be given to the type of work clothes being worn; if they become saturated with solvents, they will act as a reservoir to bathe the body continually with the harmful material.

Most volatile (easily vaporized) organic compounds are eliminated from the body in a matter of hours or, at most, days. Many of the poisonous elements, however, can be stored for long periods of time in various parts of the body. Chronic (long term) toxicity damage is unlikely to have an even distribution throughout the body. In toxicity studies with radioactive isotopes, the organ which suffers the most severe damage and appears to contribute most to the toxic effect on the body as a whole, is called the critical organ. The particular organ that shows the largest amount of damage is the one that is chosen for estimating the effect.

Industrial poisoning may be classified as either acute or chronic. The classification is based on the rate of intake of harmful materials, rate of onset of symptoms, and the duration of symptoms.

Acute poisoning is characterized by rapid absorption of the material and sudden, severe exposure. For example, inhaling high levels of carbon monoxide or swallowing a large quantity of cyanide compound will produce acute poisoning. Generally, acute poisoning results from a single dose which is rapidly absorbed and damages one or more of the vital physiological processes. The development of cancer long after recovery from acute radiation damage is called a delayed acute effect.

Chronic poisoning is characterized by absorption of a harmful material in small doses over a long period of time; each dose, if taken alone, would barely be effective. In chronic poisoning, the harmful materials remain in the tissues, continually injuring a body process. The symptoms in chronic poisoning are usually different from the symptoms seen in acute poisoning by the same toxic agent.

The Occupational Safety and Health Act of 1970 requires that the Health and Human Services publish at least annually, a list of all known toxic substances by generic family, or other useful grouping, and the concentrations at which such toxicity is known to occur. Under the OSHA Act, the Secretary of Labor must issue regulations requiring employers to monitor employee exposure to toxic materials and to keep records of any such exposure.

The purpose of The Toxic Substances List is to identify "all known toxic substances" in accordance with definitions that may be used by all sections of our society to describe toxicity. An excerpt of this list is illustrated in Figure 1. It must be emphatically stated that the presence of a substance on the list does not automatically mean that it is to be avoided. A listing does mean, however, that the substance has the documented potential of being hazardous if misused, and, therefore, care must be exercised to prevent tragic consequences.

TABLE Z-1-CONTINUED

SUBSTANCE	p/m[a]	mg./M[3b]
RONNEL		15
ROTENONE (COMMERCIAL)		5
SELENIUM COMPOUNDS (AS Se)		0.2
SELENIUM HEXAFLOURIDE	0.05	0.4
SILVER, METAL, AND SOLUBLE COMPOUNDS		0.01
SODIUM FLUOROACETATE (1080)-SKIN		0.05
SODIUM HYDROXIDE		2
STIBINE	0.1	0.5
STODDARD SOLVENT	500	2,900
STRYCHNINE		0.15
SULFUR DIOXIDE	5	13
SULFUR HEXAFLUORIDE	1,000	6,000
SULFURIC ACID		1
SULFUR MONOCHLORIDE	1	6
SULFUR PENTAFLUORIDE	0.025	0.25
SULFURYL FLUORIDE	5	20
SYSTOX, SEE DEMETON®		
2,4,5T		10
TANTALUM		5
TEDP-SKIN		0.2
TELLURIUM		0.1
TELLURIUM HEXAFLUORIDE	0.02	0.2
TEPP-SKIN		0.05
C TERPHENYLS	1	9
1,1,1,2-TETRACHLORO-2,2-DIFLUOROETHANE	500	4,170
1,1,2,2-TETRACHLORO-2,2-DIFLUOROETHANE	500	4,170
1,1,2,2-TETRACHLOROETHANE-SKIN	5	35
TETRACHLOROMETHANE, SEE CARBON TETRACHLORIDE		
TETRACHLORONAPHTHALENE-SKIN		2
TETRAETHYL LEAD (AS Pb)-SKIN		0.075
TETRAHYDROFURAN	200	590
TETRAMETHYL LEAD (AS Pb)-SKIN		0.075
TETRAMETHYL SUCCINONITRILE-SKIN	0.5	3
TETRANITROMETHANE	1	8
TETRYL (2,4,6-TRINITROPHENYL-METHYL-NITRAMINE)-SKIN		1.5
THALLIUM (SOLUBLE COMPOUNDS)-SKIN AS T1		0.1
THIRAM		5
TIN (INORGANIC CMPDS, EXCEPT OXIDES		2
TIN (ORGANIC CMPDS)		0.1
C TOLUENE-2,4-DIISOCYANATE	0.02	0.14

TABLE Z-1-CONTINUED

SUBSTANCE	p/m[a]	mg./M[3b]
O-TOLUIDINE-SKIN	5	22
TOXAPHENE SEE CHLORINATED CAMPHENE		
TRIBUTYL PHOSPHATE		5
1,1,1-TRICHLOROETHANE, SEE METHYL CHLOROFORM		
1,1,2-TRICHLOROETHANE-SKIN	10	45
TITANIUMDIOXIDE		15
TRICHLOROMETHANE, SEE CHLOROFORM		
TRICHLORANAPTHALENE-SKIN		5
1,2,3-TRICHLOROPROPANE	50	300
1,1,2-TRICHLORO 1,2,2-TRIFLUORETHANE	1,000	7,600
TRIETHYLAMINE	25	100
TRIFLUOROMONOBROMOMETHANE	1,000	6,100
2,4,6-TRINITROPHENOL SEE PICRIC ACID		
2,4,6-TRINITROPHENYLMETHYL-NITRAMINE. SEE TETRYL		
TRINITROTOLUENE-SKIN		1.5
TRIORTHOCRESYL PHOSPHATE		0.1
TRIPHENYL PHOSPHATE		3
TURPENTINE	100	560
URANIUM (SOLUBLE COMPOUNDS)		0.05
URANIUM (INSOLUBLE COMPOUNDS)		0.25
C VANADIUM		
V_2O_5 DUST		0.5
V_2O_5 FUME		0.1
VINYL BENZENE SEE STYRENE		
VINYLCYANIDE SEE ACRYLONITRILE		
VINYL TOLUENE	100	480
WARFARIN		0.1
XYLENE (XYLOL)	100	435
XYLIDINE-SKIN	5	25
YTTRIUM		1
ZINC CHLORIDE FUME		1
ZINC OXIDE FUME		5
ZIRCONIUM COMPOUNDS (AS Zr)		5

†THIS STANDARD APPLIES IN COTTON YARN MANUFACTURING UNTIL COMPLIANCE WITH § 1910.1043 (c) AND (e) IS ACHIEVED

a PARTS OF VAPOR OR GAS PER MILLION PARTS OF CONTAMINATED AIR BY VOLUME AT 25° C AND 760mm Hg PRESSURE

b APPROXIMATE MILLIGRAMS OF PARTICULATE PER CUBIC METER OF AIR
(NO FOOTNOTE "C" IS USED TO AVOID CONFUSION WITH CEILING VALUE NOTATIONS)

d AN ATMOSPHERIC CONCENTRATION OF NOT MORE THAN 0.02 P.P.M. OR PERSONAL PROTECTION MAY BE NECESSARY TO AVOID HEADACHE

TABLE Z-2

MATERIAL	8-HOUR TIME WEIGHTED AVERAGE	ACCEPTABLE CEILING CONCENTRATION	ACCEPTABLE MAXIMUM PEAK ABOVE THE ACCEPTANCE CEILING CONCENTRATION FOR AN 8-HOUR SHIFT	
			CONCENTRATION	MAXIMUM DURATION
BENZENE (Z37.40-1969)	10 P.P.M.	25 P.P.M.	50 P.P.M.	10 MINUTES
BERYLLIUM AND BERYLLIUM COMPOUNDS (Z37.29-1970)	2μ G./M³	5μ G./M³	25μ G./M³	30 MINUTES
CADMIUM FUME (Z37.5-1970)	0.1 MG/M³	0.3 MG/M³		
CADMIUM DUST (Z37.5-1970)	0.2 MG/M³	0.6 MG/M³		
CARBON DISULFIDE (Z37.3-1968)	20 P.P.M.	30 P.P.M.	100 P.P.M.	30 MINUTES
CARBON TETRACHLORIDE (Z37.17-1967)	10 P.P.M.	25 P.P.M.	200 P.P.M.	5 MINUTES IN ANY 4 HOURS
CHROMIC ACID AND CHROMATES (Z37.7-1971)		1 MG/10M³		

Figure 1 Excerpt of Toxic Substance List

The absence of a substance from the list does not necessarily indicate that a substance is not toxic. Some hazardous substances may not qualify for the list because the dose that causes the toxic effect is not known.

Other chemicals associated with skin sensitization and carcinogenicity (ability to cause cancer) may be omitted from the list, because these effects have not been reproduced in experimental animals or because the human data is not definitive.

It is not the purpose of the list to quantify the hazard by way of the toxic concentration or dose that is presented with each of the substances listed. Hazard evaluation involves far more than the recognition of a toxic substance and a knowledge of its relative toxic potency. It involves a measurement of the quantity that is available for absorption by the user, the amount of time that is available for absorption, the frequency with which the exposure occurs, the physical form of the substances, and the presence of other substances, additives, or contaminants (toxic or non-toxic).

The purpose of the Material Safety Data Sheet (MSDS) is to ensure the individuals working with chemicals and in the vicinity of chemicals have specific information on these chemicals. This form identifies the chemical by its technical and common name and lists the physical/chemical characteristics and fire, explosion, and reactivity hazards. The second page specifies health hazards and recommends first aid procedures. The safe handling and control measures are also supplied. The MSDS is a very helpful document, and personnel working around chemicals should make it a practice to review these sheets frequently for their own safety. Figures 2 and 3 are copies of the MSDS.

The Code of Federal Regulations recommends that the hazards of all chemicals produced and imported be evaluated and the information concerning the hazards be transmitted to the employers and employees. The MSDS, labels on containers, and employee training should be part of a comprehensive hazards communication program.

Material Safety Data Sheet May be used to comply with OSHA's Hazard Communication Standard, 29 CFR 1910.1200. Standard must be consulted for specific requirements.	**U.S. Department of Labor** Occupational Safety and Health Administration (Non-Mandatory Form) Form Approved OMB No. 1218-0072
IDENTITY *(As Used on Label and List)*	*Note: Blank spaces are not permitted. If any item is not applicable, or no information is available, the space must be marked to indicate that.*

Section I

Manufacturer's Name	Emergency Telephone Number
ADDRESS *(Number, Street, City, State, and ZIP Code)*	Telephone Number for Information
	Date Prepared
	Signature of Preparer (optional)

Section II – Hazardous Ingredients/Identity Information

Hazardous Components *(Specify Chemical Identify: Common Name(s))*	OSHA PEL	ACGIH TLV	Other Limits Recommended	*% (optional)*

Section III – Physical/Chemical Characteristics

Boiling Point		Specific Gravity (H_2O=1)	
Vapor Pressure (mm Hg.)		Melting Point	
Vapor Density (AIR=1)		Evaporation Rate (Butyl Acetate=1)	
Solubility in Water			
Appearance and Odor			

Section IV – Fire and Explosion Hazard Data

Flash Point (Method Used)	Flammable Limits	LEL	UEL
Extinguishing Media			
Special Fire Fighting Procedures			
Unusual Fire and Explosion Hazards			

(Reproduce locally) OSHA 174. Sept. 1985

Figure 2 Typical Material Safety Data Sheet (Sections I-IV)

Section V - Reactivity Data

Stability	Unstable		Conditions to Avoid
	Stable		

Incompatibility (Materials to Avoid)

Hazardous Decomposition or Byproducts

Hazardous Polymerization	May Occur		Conditions to Avoid
	Will Not Occur		

Section VI - Health Hazard Data

Route(s) of Entry: Inhalation? Skin? Ingestion?

Health Hazards *(Acute and Chronic)*

Carcinogenicity: NTP? IARC Monographs? OSHA Regulated?

Signs and Symptoms of Exposure

Medical Conditions Generally Aggravated by Exposure

Emergency and First Aid Procedures

Section VII - Precautions for Safe Handling and Use

Steps to Be Taken in Case Material is Released or Spilled

Waste Disposal Method

Precautions to Be Taken in Handling and Storing

Other Precautions

Section VIII - Control Measures

Respiratory Protection *(Specify Type)*

Ventilation	Local Exhaust	Special
	Mechanical (General)	Other

Protective Gloves	Eye Protection

Other Protective Clothing or Equipment

Work/Hygienic Practices

Page 2 ✱ USGPO 1986-491-529/45775

Figure 3 Typical Material Safety Data Sheet (Sections V-VIII)

Summary

The important information in this chapter is summarized below.

Toxic Compounds Summary

- The general safety precautions regarding toxic compounds:

  Proper ventilation
  Appropriate hygienic practices
  Housekeeping
  Protective clothing
  Training

- The criteria used to determine if a compound is a health hazard:

  Toxicity of the materials used
  The physical properties
  The absorption probabilities of these materials by individuals
  The extent and intensity of exposure to these materials
  The control measures used

- The methods by which toxic compounds may enter the body:

  Ingestion
  Inhalation
  Absorption

- The purpose and general contents of the Material Safety Data Sheets (MSDS) is to ensure the individuals working with and in the vicinity of chemicals have specific information of these chemicals. This form identifies the chemical, by technical and common name, lists the physical/chemical characteristics, any fire or explosion hazard as well as reactivity hazards. The second page will specify health hazards and recommend first aid procedures. The safe handling and control measures are also supplied.

- The purpose and general contents of the Toxic Substance List is to identify "all known toxic substances" in accordance with definitions that may be used by all sections of our society to describe toxicity. This form identifies known toxic chemicals which have been proven in lab tests or have definite human data. It lists the chemical name, level of concentration at which it is hazardous, concentration limits for set time exposures both weighted averages and ceiling limits.

COMPRESSED GASES

Gases are commonly used throughout industry. These gases come in several forms and are often as dangerous as they are useful. This chapter provides background knowledge of these gases.

EO 1.7 DEFINE the following terms:

a. Compressed gas
b. Non-liquified gases
c. Liquified gases
d. Dissolved gases

EO 1.8 STATE the five major families of gases.

EO 1.9 STATE the general safety precautions regarding the use, handling, and storage of gases.

EO 1.10 STATE the safety precautions for working with cryogenic liquids.

EO 1.11 LIST the physical properties and special precautions for the following gases:

a. Hydrogen
b. Oxygen
c. Nitrogen

Compressed Gases

Gases in compressed form serve countless indispensable roles in modern technology. Oxygen is used extensively to produce stronger and cheaper steels. Acetylene welding and brazing of certain metals has been common for many years. Other compressed, flammable gases such as hydrogen are equally necessary for the welding of certain metals. Some metals and alloys (such as stainless steel, titanium, and zirconium) can be welded only under an inert gas atmosphere. Carbon dioxide is used extensively in fire extinguishers for chemical and electrical fires. In the nuclear industry, uses of compressed gases range from the addition of nuclear grade hydrogen to reactor plant systems to propane and butane for heating components or spaces.

Gases are compressed for practical reasons of transportation, storage and use. The definition of *compressed gas* by the Interstate Commerce Commission (ICC) reads: "... any material or mixture having in the container an absolute pressure exceeding 40 psi (pounds per square inch) at 70°F, or regardless of the pressure at 70°F, having an absolute pressure exceeding 140 psi at 130°F; or any flammable material having a vapor pressure exceeding 40 psi at 100°F."

Because we often deal in gage pressures, and absolute pressure is equal to atmospheric pressure (14.7 psi at sea level) plus the pressure that would be read on an ordinary gage, we can simplify the above definition. Accordingly, a compressed gas is one that gives a pressure reading of:

1. either 25 psig (pounds per square inch gage) at 70°F; or 125 psig at 130°F; or

2. if the contained substance is flammable, 25 psig at 100°F.

This simplified definition and the range of boiling points among gases classify gases into two major groups that differ in physical state when contained.

1. *Non-liquified gases*, which are gases that do not liquify in containers at ambient temperatures and under pressures attained in commercially used containers that range to 2000 to 2500 psig.

2. *Liquified gases*, which are gases that do become liquids to a very large extent in containers at ordinary ambient temperatures and at pressures from 25 to 2500 psig.

The first group, commonly called non-liquified gases, have relatively low boiling points, approximately -150°F or lower.

Non-liquified gases do, however, become liquids if cooled to temperatures below their boiling points. Those that liquify at "cryogenic" temperatures (from absolute zero [-459.7°F] to around -240°F) are known as cryogenic fluids.

The second group, or liquified gases, have boiling points relatively near atmospheric temperatures (from about -130°F to 30°F). The liquified gases solidify at cryogenic temperatures.

Oxygen, helium and nitrogen are examples of gases in wide use both as non-liquified gases and cryogenic fluids. With respective boiling points of -297°F, -425°F, and -320°F, they are charged into high pressure steel cylinders at more than 2000 psig at 70°F for shipment and use as non-liquified gases. However, when shipped as cryogenic fluids, they are cooled down to liquid form and charged into special insulated containers that keep them below their boiling points and are contained at pressures normally less than 75 psig.

A third physical state in the container is represented by only one widely used gas, acetylene. Acetylene is sometimes referred to as a *dissolved gas*. A dissolved gas is defined as a gas that is dissolved into a solution.

The industry recommends that free acetylene should not ordinarily be handled at pressures greater than 15 psig because, if handled at higher pressures without special equipment, it can decompose with explosive violence. Consequently, acetylene cylinders are packed with an inert porous material that is saturated with acetone. Acetylene charged into the cylinder dissolves in the acetone and in solution will not decompose at or below the maximum authorized shipping pressure of 250 psig at 70°F.

Compressed or liquified gases are also often described according to loosely-knit families to which they belong through common origins, properties, or uses. The major families of gases are atmospheric gases, fuel gases, refrigerant gases, aerosol gases, and poison gases.

Atmospheric gases comprise one family. Its most abundant member is nitrogen, constituting 78 percent of air by volume; oxygen, constituting 21 percent of air by volume, is its second most abundant member. Most of the remaining 1 percent of the atmosphere consists of a sub-family of gases, the inert gases, that share the property of chemical inertness. Inert gases are chiefly argon, with minute quantities of helium, neon, krypton, xenon and radon. The last four are frequently called the rare gases due to their scarcity. Hydrogen also occurs minutely in the atmosphere, as do a large variety of trace constituents, small amounts of carbon dioxide, and large amounts of water vapor.

Another family of gases are the fuel gases. Fuel gases burned in air or with oxygen to produce heat make up a large family related through their major use. Its members are notably the hydrocarbons including liquefied petroleum (LP) gases, propane, butane, methane, and welding gases such as acetylene and hydrogen.

An opposite application relates members of another large family, the refrigerant gases. A refrigerant gas liquifies easily under pressure and works by being compressed to a liquid which then absorbs large amounts of heat as it circulates through coils where it vaporizes back into gaseous form. Examples of refrigerant gases include ammonia and the fluorocarbons (freon).

Aerosol propellant gases make up a family also related by use through the introduction of pressure-packaged products used in the form of a spray or a foam. Propellant gases have moderate vapor pressures at room temperatures (70 psig down to 35 psig, and even lower in some cases). It is usually agreed that a good propellant should also be nontoxic, chemically stable, noncorrosive, and inexpensive. The fluorocarbons (freon) and nitrous oxide are the most commonly used propellant gases.

Gases considered to be members of the poison gas family are generally those that the ICC has classified as poison gases to ensure public safety in interstate shipments. Two examples of these gases are hydrogen cyanide and phosgene.

Basic Safety Precautions Regarding Compressed Gases

Compressed and liquified gases are widely useful due to properties including high heat output in combustion for some gases, high reactivity in chemical processing with other gases, extremely low temperatures available from some gases, and the economy of handling them all in compact form at high pressure or low temperature. These same properties, however, also represent hazards if the gases are not handled with full knowledge and care.

Practically all gases can act as simple asphyxiants by displacing the oxygen in air. The chief precaution taken against this potential hazard is adequate ventilation of all enclosed areas in which unsafe concentrations may build up. A second precaution is to avoid entering unventilated areas that might contain high concentrations of gas without first putting on breathing apparatus with a self-contained or hose-line air supply. A number of gases have characteristic odors which can warn of their presence in air. Others, however, like the atmospheric gases, have no odor or color. Warning labels are required for compressed and liquified gas shipping containers. Similar warning signs are placed at the approaches to areas in which the gases are regularly stored and used.

Some gases can also have a toxic effect on the human system, either inhalation, through high vapor concentrations, or by liquified gas coming in contact with the skin or the eyes. Adequate ventilation of enclosed areas serves as the chief precaution against high concentrations of gas. In addition, for unusually toxic gases, automatic devices can be purchased or built to monitor the gas concentration constantly and set off alarms if the concentration approaches a danger point. Precautions against skin or eye contact with liquified gases that are toxic or very cold, or both, include thorough knowledge and training for all personnel handling such gases, the development of proper procedures and equipment for handling them, and special protective clothing and equipment (for example, protective garments, gloves, and face shields).

With flammable gases, it is necessary to guard against the possibility of fire or explosion. Ventilation, in addition to safe procedures and equipment to detect possible leaks, represents a primary precaution against these hazards. If fire breaks out, suitable fire extinguishing apparatus and preparation will limit damage. Care must also taken to keep any flammable gas from reaching any source of ignition or heat (such as sparking electrical equipment, sparks struck by ordinary tools, boiler rooms, or open flames).

Oxygen poses a combustible hazard of a special kind. Although oxygen does not ignite, it lowers the ignition point of flammable substances and greatly accelerates combustion. It should not be allowed closer than 10 feet to any flammable substance, including grease and oil, and should be stored no closer than 10 feet to cylinders or tanks containing flammable gases.

Proper storage and handling of containers avoids many possible incidents. Hazards resulting from the rupture of a cylinder or other vessel containing gas at high pressure are protected against by careful and secure handling of containers at all times. For example, cylinders should never be struck nor allowed to fall, because if the cylinder is charged to a high pressure and the cylinder valve is broken off, it could become a projectile. Cylinders should not be dragged or rolled across the floor; they should be moved by a hand truck. Also, when they are upright on a hand truck, floor, or vehicle, they should be chained securely to keep them from falling over. Moreover, cylinders should not be heated to the point at which any part of their outside surface exceeds a temperature of 125°F, and they should never be heated with a torch or other open flame. Similar precautions are taken with larger shipping and storage containers. Initial protection against the possibility of vessel rupture is provided by the demanding requirements and recommendations that compressed gas containers fulfill in their construction, testing and retesting.

Cryogenic Liquids

Most cryogenic liquids are colorless, odorless, and tasteless when vaporized to a gas. As liquids, most have no color (except liquid oxygen which is light blue). However, whenever the cold liquid and vapor are exposed to the atmosphere a warning appears. As the boil-off gases condense moisture in the air, a fog forms that extends over an area larger than the vaporizing gas. Many cryogenic liquids are inert gases, and may inert an enclosed space. Inert gases will not support life.

Both the liquid and its boil-off vapor can rapidly freeze human tissue and can cause many common materials such as carbon steel, plastic, and rubber to become brittle or fracture under stress. Liquids in containers and piping at temperatures at or below the boiling point of liquified air (-318°F) can cause the surrounding air to condense to a liquid.

Extremely cold liquified gases (helium, hydrogen, and neon) can even solidify air or other gases to which they are directly exposed. In some cases, plugs of ice or foreign material will develop in cryogenic container vents and openings and cause the vessel to rupture. If a plug forms, contact the supplier immediately. Do not attempt to remove the plug; move the vessel to a remote location.

All cryogenic liquids produce large volumes of gas when they vaporize. For example, 1 volume of saturated liquid nitrogen at 1 atmosphere vaporizes to 696.5 volumes of nitrogen gas at room temperature at 1 atmosphere.

When vaporized in a sealed container, cryogenic liquids produce enormous pressures. If 1 volume of liquid helium at 1 atmosphere is warmed to room temperature and vaporized in a totally enclosed container, it has the potential to generate a pressure of more than 14,500 psig. Because of this high pressure, cryogenic containers are usually protected with two pressure-relief devices, a pressure-relief valve and a frangible (easily broken) disk.

Many safety precautions that must be taken with compressed gases also apply to liquified gases. However, some additional precautions are necessary because of the special properties exhibited by fluids at cryogenic temperatures.

The properties of cryogenic liquids affect their safe handling and use. Table 1 presents information to help determine safe handling procedures. None of the gases listed are corrosive at ambient temperatures, and only carbon monoxide is toxic.

- Always handle cryogenic liquids carefully. They can cause frostbite on skin and exposed eye tissue. When spilled, they tend to spread, covering a surface completely and cooling a large area. The vapors emitted by these liquids are also extremely cold and can damage tissues. The vapor boil-off may inert the immediate vicinity.

- Stand clear of boiling or splashing liquid and its vapors. Boiling and splashing occurs when a warm container is charged or when warm objects are inserted into a liquid. These operations should always be performed slowly to minimize boiling and splashing. If cold liquid or vapor comes in contact with the skin or eyes, first aid should be given immediately.

- Never allow an unprotected part of the body to touch uninsulated pipes or vessels that contain cryogenic fluids. The extremely cold metal will cause the flesh to stick fast to the surface and tear when withdrawn. Touching even nonmetallic materials at low temperatures is dangerous.

Tongs, or a similar device, should be used to withdraw objects immersed in a cryogenic liquid. Materials that are soft and pliable at room temperature become hard and brittle at extremely low temperatures and will break easily.

Workers handling cryogenic liquids should use eye and hand protection to protect against splashing and cold-contact burns. Safety glasses are also recommended. If severe spraying or splashing is likely, a face shield or chemical goggles should be worn. Protective gloves should always be worn when anything that comes in contact with cold liquids and their vapors is being handled. Gloves should be loose fitting so that they can be removed quickly if liquids are spilled into them. Trousers should remain outside of boots or work shoes.

TABLE 1
Safety Properties of Cryogenic Fluids

	Xenon (Xe)	Krypton (Kr)	Methane (CH_4)	Oxygen (O_2)	Argon (Ar)	Carbon Monoxide (CO)	Nitrogen (N_2)	Neon (Ne)	Hydrogen (H_2)	Helium (He)
Boiling Point, 1 atm °F °C	-163 -108	-244 -153	-259 -161	-297 -183	-303 -186	-313 -192	-321 -196	-411 -246	-423 -253	-425 -268
Melting Point, 1 atm °F °C	-169 -112	-251 -157	-296 -182	-362 -219	-309 -189	-341 -207	-346 -210	-416 -249	-435 -259	N/A
Density, boiling point, 1 atm lb/cu ft	191	151	26	71	87	49	50	75	4.4	7.8
Heat of vaporization boiling point Btu/lb	41	46	219	92	70	98	85	37	193	10
Volume expansion ratio, liquid at 1 atm boiling point to gas at 60° F, 1 atm	559	693	625	881	841	N/A	697	1447	850	754
Flammable	No	No	Yes	N/A	No	Yes	No	No	Yes	No

Treating Cold-Contact Burns

Workers will rarely come in contact with a cryogenic liquid if proper handling procedures are used. In the unlikely event of contact with a liquid or cold gas, a cold-contact "burn" may occur. Actually, the skin or tissue freezes. Medical assistance should be obtained as soon as possible. In the interim, the emergency measures presented in Table 2 are recommended.

TABLE 2
Emergency Measures for Treating Cold-Contact Burns

- Remove any clothing that may restrict circulation to the frozen area. Do not rub frozen parts, as tissue damage may result.
- As soon as practical, immerse the effected part in warm water (not less than 105°F or more than 115°F, or 40°C to 46°C). Never use dry heat. The victim should be in a warm room, if possible.
- If the exposure has been massive and the general body temperature is depressed, the patient should be totally immersed in a warm-water bath. Treatment for shock should be provided.
- Frozen tissues are painless and appear waxy and yellow. They will swell and be painful and prone to infection when thawed. Do not rewarm rapidly. Thawing may require 15 to 60 minutes and should continue until the pale blue tint of the skin turns pink or red. Morphine or tranquilizers may be required to control the pain during thawing and should be administered under professional medical supervision.
- If the frozen part of the body thaws before the doctor arrives, cover the area with dry sterile dressings and a large, bulky protective covering.
- Alcoholic beverages and smoking decrease blood flow to the frozen tissues and should be prohibited. Warm drinks and food may be administered.

Some liquified gases require special precautions. For example, when liquid oxygen is handled, all combustible materials, especially oil or gases, should be kept away. Smoking or open flames should never be permitted where liquid oxygen is stored or handled. NO SMOKING signs should be posted conspicuously in such areas.

Liquid oxygen, or oxygen-rich air atmospheres, should not come in contact with organic materials or flammable substances. Some organic materials (oil, asphalt, kerosene, cloth, or dirt containing oil or grease) react violently with oxygen, and may be ignited by a hot spark. Liquid oxygen may form mixtures that are shock sensitive with fuels, oils, or grease. If liquid oxygen spills on asphalt, or on another surface contaminated with combustibles (for example, oil-soaked concrete or gravel), no one should walk on, and no equipment should pass over, the area for at least 30 minutes after all frost or fog has disappeared.

Any clothing that has been splashed or soaked with liquid oxygen, or exposed to a high gaseous-oxygen atmosphere, should be changed immediately. The contaminated systems should be aired for at least an hour so that they are completely free of excess oxygen. Workers exposed to high-oxygen atmospheres should leave the area and avoid all sources of ignition until the clothing and the exposed area have been completely ventilated. Clothing saturated with oxygen is readily ignitable and will burn vigorously.

Specific Properties of Selected Industrial Gases

Anyone who uses gases must have a thorough knowledge of their chemical properties to maintain a controlled operation. If the gas is flammable, its flammable range and ignition temperature must be known.

The lower flammable limit is the smallest percent of the gas in air which can ignite when exposed to the ignition temperature. The upper flammable limit is the point above which the mixture is too rich in fuel to ignite. The range between these two limits is the flammable, or explosive, range. The most violent explosion will occur at concentrations about the middle of the flammable range. Sources of heat that may cause temperatures that exceed the ignition temperature must be avoided, as well as gas-air mixtures that are within the flammable range.

The physiological effects of the gas must be known, not only types of reactions, but also severity of reactions. All employees who handle gas should be familiar with its effects, and recommended control measures.

The chemical reactivity of the gas must be known. This includes a knowledge of the materials that are resistant to its chemical effects, the materials with which it reacts, and how it reacts with such materials. Some gases become unstable at high pressures, and others become more corrosive at high temperatures.

The term Threshold Limit Value (TLV) is sometimes used and is defined as: The highest time-weighted average concentration of an air contaminant which if breathed for a normal working day is unlikely to result in health injury to the average person, either at the time, or after years of exposure.

The following discussion addresses some of the more commonly used gases, and describes the specific chemical properties and characteristics that are important for accident prevention.

Hydrogen

Hydrogen (H_2) is the lightest of all elements. Its presence cannot be detected by any of the senses. It is flammable in oxygen or air, and has a flammable range of from 4.1 percent to 74.2 percent by volume in air. A mixture of 10 to 65 percent hydrogen by volume in air will explode if ignited. Pure hydrogen burns quietly in air with an almost invisible flame, and when burned with pure oxygen, a very high temperature may be reached. Hydrogen will burn readily in chlorine gas, and under proper conditions, will combine with nitrogen, forming ammonia.

Some chemical reactions produce hydrogen as a byproduct. A lead-acid battery will produce hydrogen when it is being charged. Metallic sodium and potassium are examples of some chemicals that react violently when exposed to water, producing hydrogen, which may flame spontaneously due to the heat of the reaction. Many electroplating processes produce hydrogen. Some chemicals used to remove scale from the water side of boilers give off hydrogen. Whatever the operation, it is important to know whether hydrogen will be produced, and if so, precautions must be taken to prevent its accumulation and ignition. The precautions to take include adequate ventilation to prevent its accumulation and the elimination of possible sources of ignition. Hydrogen is classified as an asphyxiant.

Nitrogen

Nitrogen (N_2) makes up more than 78 percent of the earth's atmosphere. It will not burn or support combustion. It cannot be detected by any of the senses and it is not toxic. Although it is often referred to as an inert gas because it does not oxidize readily, it nevertheless forms many compounds. It is frequently used to inert systems that contain, or have contained, flammable liquids or gases. Inerting a system means replacing the oxygen with an inert gas in order to reduce the possibility of fire or explosion.

Nitrogen is fairly soluble in the blood, and a considerable amount will dissolve in the blood of a person when the air pressure is increased, as in diving, caisson, and some tunnel work. If these employees are not properly decompressed, the dissolved nitrogen escapes from the blood in the form of small bubbles in the bloodstream causing intense pain and is often fatal. This disorder is commonly known as the bends.

If a large amount of nitrogen were released into the air of an enclosed space, it could cause a serious oxygen deficiency. Nitrogen is an asphyxiant.

Oxygen

Oxygen (O_2) supports combustion, but does not burn. Even so, it must be considered a potentially hazardous element from a fire hazard standpoint. The results of an enriched oxygen atmosphere include a lowered ignition temperature, an increased flammable range, and an acceleration of the burning rate. Oxygen readily combines with other elements and compounds, with spontaneous ignition in some cases. When oxygen comes in contact with oil, grease, or fuel oils, it may ignite violently. Every possible precaution must be taken to prevent this combination.

Oxygen sustains life, but if pure oxygen were inhaled continuously for extended periods, the reactions in the body would be too rapid and would cause harmful effects. Oxygen should always be referred to as oxygen, and not air, to prevent confusion. It should never be used to run pneumatic equipment because of the possibility of coming in contact with oil that may be inside the equipment. Finally, oxygen valves should be operated slowly. Abruptly starting and stopping oxygen flow may ignite contaminants in the system.

Sources of Ignition

All known sources of ignition must be eliminated in areas that contain, or may contain, flammable gases. This includes areas where gases are being manufactured or used in some process and in areas where they are stored.

One potential source of ignition is electrical equipment. When used or installed in hazardous locations, this equipment must be explosion-proof and properly installed.

Electrical equipment includes not only the more obvious equipment such as motors, generators, motor controls, switches, and lighting fixtures, but also the not so obvious equipment such as alarm systems, remote controls, telephones and other communication systems.

The use of unapproved portable electric tools and equipment should be strictly prohibited in hazardous locations. When temporary lighting is used, it must be an approved type and in excellent repair.

All possible sources of static electricity should be anticipated to prevent its buildup and discharge. Several methods of control may be used. It is necessary that conductive parts of a system be bonded (described in the following chapter) together to eliminate the difference in potential between the parts, and the whole system grounded to eliminate the difference in potential between the system and ground.

When some gases come in contact with certain other substances, spontaneous ignition may occur. One of the better known of these reactions is that of oxygen with oil. Another reaction is the possible ignition of hydrogen due to the heat of the reaction which liberated it, such as sodium in water. Additional sources of ignition are as follows.

- The spontaneous ignition that may occur in oily rags and other materials.
- Smoking is a principal cause of fire, and calls for strict prohibition in hazardous areas.
- Other sources of ignition include sparks or flame-producing operations such as grinding or welding, and equipment such as heaters or boilers.
- Ferrous handtools can produce a spark when struck against, or by a suitable object, and may cause ignition of some gas-air mixtures.
- Other possible sources of ignition are internal combustion engines and battery-powered electric vehicles.

The most disastrous accidents involving industrial gases have been caused by ignition of gas-air mixtures. Fires cannot occur without three essential factors being present in the right quantity, at one place, and at the same time. These factors are fuel, heat, and oxygen (which is normally supplied by air). When they are present in the same place, in the right amounts, and at the same time, there will be a fire. To prevent a fire, it is necessary to keep only one of these factors away from the other two. Air is necessary for life and cannot normally be eliminated, but the fuel and heat can be kept separated. Proper operational and maintenance procedures will help prevent a flammable gas from escaping into the air, but because this is always a possibility, it is most important to eliminate the source of ignition.

Summary

The important information in this chapter is summarized below.

Compressed Gases Summary

- The following terms are defined:

  Compressed gas is defined as any material or mixture having in the container an absolute pressure exceeding 40 psi (pounds per square inch) at 70°F, or regardless of the pressure at 70°F, having an absolute pressure exceeding 140 psi at 130°F; or any flammable material having a vapor pressure exceeding 40 psi at 100°F.

Compressed Gases Summary (Cont.)

Non-liquified gases are gases that do not liquify in containers at ambient temperatures and under pressures attained in commercially used containers that range to 2000 to 2500 psig.

Liquified gases are gases which do become liquids to a very large extent in containers at ordinary ambient temperatures, and at pressures from 25 to 2500 psig.

Dissolved gases are gases dissolved into a solution.

- The major families of gases are:

  atmospheric gases
  fuel gases
  refrigerant gases
  aerosol gases
  poison gases

- The general safety precautions for gases:

  Ensure there is adequate ventilation of enclosed areas.

  Avoid entering unventilated areas that might contain high concentrations of gas without first putting on breathing apparatus with a self-contained or hose-line air supply.

  Take precautions against skin or eye contact with liquified gases that are toxic or very cold, or both.

  The proper storage and handling of containers avoids many possible incidents.

- The safety precautions for working with cryogenic liquids are:

  Stand clear of boiling or splashing liquid and its vapors.

  Never allow an unprotected part of the body to touch uninsulated pipes or vessels that contain cryogenic fluids.

  Use tongs, or a similar device, to withdraw objects immersed in a cryogenic liquid.

  Use appropriate eye and hand protection.

Compressed Gases Summary (Cont.)

- The physical properties and special precautions for hydrogen include the following:

  It is the lightest of all elements
  It cannot be detected by any of the senses
  It has a flammable range of from 4.1 percent to 74.2 percent by volume in air
  It burns with an almost invisible flame

- Hydrogen is classified as an asphyxiant, and special precautions, such as adequate ventilation and the elimination of possible sources of ignition, should be taken to prevent hydrogen accumulation.

- The physical properties and special precautions for nitrogen include the following:

  It makes up more than 78 percent of the earth's atmosphere
  It will not burn or support combustion
  It cannot be detected by any of the senses
  It is used to inert systems that contain, or have contained, flammable liquids or gases
  It is fairly soluble in the blood and if the dissolved nitrogen escapes from the blood in the form of small bubbles in the bloodstream, intense pain occurs and effects are often fatal

- Nitrogen is an asphyxiant and it could cause a serious oxygen deficiency. Special precautions, such as adequate ventilation, should be taken to prevent nitrogen accumulation.

- The physical properties and special precautions for oxygen include the following:

  It supports combustion, but does not burn
  Ignition temperature of a combustible material is lower in enriched oxygen
  It readily combines with other elements and compounds, with spontaneous ignition in some cases

- When oxygen comes in contact with oil, grease, or fuel oils, it may ignite violently. Every possible precaution must be taken to prevent this combination. If pure oxygen were inhaled continuously for extended periods the reactions in the body would be too rapid and cause harmful effects. Finally, oxygen valves should be operated slowly. Abruptly starting and stopping oxygen flow may ignite contaminants in the system.

FLAMMABLE AND COMBUSTIBLE LIQUIDS

One of the most devastating industrial accidents is an explosion resulting in fire. Carelessness is often the root cause. Improper handling and storage can also lead to disastrous results. This chapter introduces precautions taken to avoid accidents with flammable and combustible liquids.

EO 1.12 **DEFINE the following terms:**

- **a. Flammable liquid**
- **b. Combustible liquid**

EO 1.13 **STATE general safety precautions regarding the use, handling, and storage of flammable and combustible liquids.**

EO 1.14 **STATE the reasons for and techniques used in bonding and grounding of flammable liquid containers.**

EO 1.15 **LIST four sources of ignition of flammable liquids.**

EO 1.16 **STATE the health hazards associated with flammable and/or combustible liquids.**

Flammable And Combustible Liquids Definitions

As defined by OSHA, DOT, and the National Fire Protection Association Standard, Flammable and Combustible Liquids Code, NFPA 30, a *flammable liquid* is one having a flash point below 100°F, and having a vapor pressure not exceeding 40 psia at 100°F. *Combustible liquids* are those with flash points at or above 100°F, but below 200°F. Although combustible liquids do not ignite as easily as flammable liquids, they must be handled with caution because of their ability to ignite under certain conditions. Flammable and combustible liquids are further subdivided by NFPA 30 into classes as follows.

- Class I - Those liquids having flash points below 100°F.

  IA - Those liquids having flash points below 73°F and having a boiling point below 100°F.

  IB - Those liquids having flash points below 73°F and having a boiling point at or above 100°F.

IC - Those liquids having flash points at or above 73°F and below 100°F.

- Class II - Those liquids having flash points at or above 100°F and below 140°F.
- Class III - Those liquids having flash points above 140°F.

IIIA - Those liquids having flash points at or above 140°F and below 200°F.

IIIB - Those liquids having flash points above 200°F.

Flammable and combustible liquids vaporize to form flammable mixtures when they are stored in open containers, when they leak or are spilled, or when heated. The hazard that exists from these mixtures is largely dependent upon the flash point of the liquid, the concentration of the mixture, and the potential of a source of ignition at a temperature sufficient to cause the mixture to burst into flame.

Since it is the vapor-air mixture formed from the evaporation of the liquid that poses the hazard, exposures of large liquid surface areas and sources of heat should be avoided or prevented during handling or storage of these liquids.

Some of the commonly used terms associated with flammable and combustible liquids are provided below.

- Auto-Ignition Temperature -
  The minimum temperature at which a flammable mixture will ignite from its own heat source or contact with a heated surface without necessity of a spark or flame.

- Flash Point -
  The minimum temperature at which a liquid gives off vapor in sufficient concentration to form an ignitible mixture with air near the surface of the liquid (under controlled test conditions).

- Flammable Limits -
  The minimum concentration of vapor in air below which propagation of flame does not occur on contact with a source of ignition. This is known as the lower flammable limit (LFL). There is also a maximum concentration of vapor or gas in air above which propagation of flame will not occur. This is called the upper flammable limit (UFL).

- Flammable Range -
  The difference between the lower and upper flammable limits, expressed in percentage of vapor or gas in air by volume. Also known as the explosive range.

- Propagation of Flame -
  The spread of flame through the entire volume of the flammable mixture from a single source of ignition.

- Volatility -
  The tendency or ability of a liquid to vaporize.

- Oxygen Limits -
  The concentration of oxygen below which a normally flammable vapor mixture will not ignite (normally 12 to 14 percent oxygen). An increase of pressure or temperature will reduce the required lower oxygen limit, however.

- Rate of Diffusion -
  The tendency of one gas or vapor to disperse into or mix with another gas or vapor. Dependent upon the density of the vapor or gas as compared to air.

- Vapor Pressure -
  The pressure exerted by a volatile liquid under any of the conditions of equilibrium that may exist between the vapors and the liquid.

- Oxygen Deficiency -
  The designation given to an atmosphere containing less than the percentage of oxygen normally found in air (<21%).

The volatility of liquids will increase as heat is applied to them. Accordingly, combustible liquids should be treated as flammable when they are heated even though the liquid, when not heated, is outside the flammable range.

The Department of Transportation (DOT) has modified the definition of flammable liquids in its Hazardous Materials Regulation, 49 CFR parts 170 to 179. They are defined as any liquid that gives off flammable vapors at or below a temperature of 80°F. This definition is important because the DOT Flammable Liquid Label, illustrated in Figure 4, is one means by which containers of flammable liquids can be identified for shipping, receiving, and transportation.

Figure 4 DOT Flammable Liquid Labels

Safety Precautions

Avoid accidental mixture of flammable and combustible liquids. A small amount of a highly volatile substance may lower the flash point of a less volatile substance and form a more flammable mixture. In addition, the lower flash point liquid can act as a fuse to ignite the higher flash point material in the same manner as if it were a flammable mixture.

Fill and discharge lines and openings, as well as control valves associated with flammable and combustible systems, shall be identified by labels, color coding, or both, to prevent mixing different substances. All storage tanks shall be clearly labeled with the name of its contents, and products stored within shall not be intermixed. Transfer lines from different types and classes of flammable products should be kept separate, and preferably, different pumps should be provided for individual products.

For handling quantities of flammable liquids up to five gallons, a portable FM (Factory Mutual Engineering Corp.) or UL (Underwriters Laboratory) approved container should be used. The container should be clearly identified by lettering or color code.

Smoking, the carrying of strike-anywhere matches, lighters, and other spark-producing devices should not be permitted in a building or area where flammable liquids are stored, handled, or used. The extent of the restricted area will depend on the type of products handled, the design of the building, local codes, and local conditions.

Suitable NO SMOKING signs should be posted conspicuously in those buildings and areas where smoking is prohibited.

Static electricity is generated by the contact and separation of dissimilar material. For example, static electricity is generated when a fluid flows through a pipe or from an orifice into a tank. Examples of several methods of generating static electricity are shown in Figure 5. The principal hazards created by static electricity are fire and explosion, which are caused by spark discharges.

A point of great danger from a static spark is where a flammable vapor is present in the air, such as the outlet of a flammable liquid fill pipe, at a delivery hose nozzle, near an open flammable liquid container, and around a tank truck fill opening. In the presence of a mechanism for generating a static charge, a spark between two bodies occurs when there is a poor electrical conductive path between them. Hence, grounding or bonding of flammable liquid containers is necessary to prevent static electricity from causing a spark.

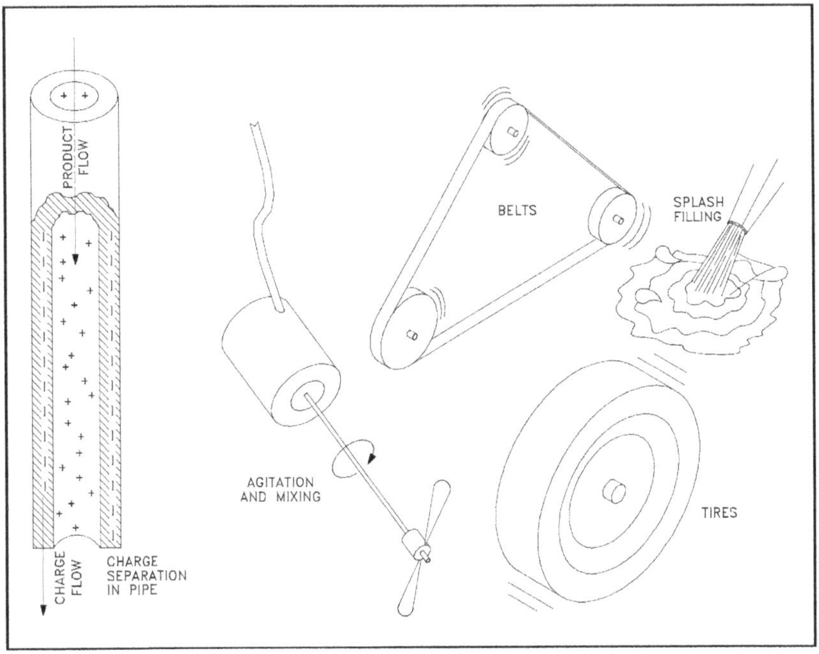

Figure 5 Typical Static-producing Situations, Including Charge Separation in Pipe

The terms bonding and grounding have sometimes been used interchangeably because of a poor understanding of the terms. As illustrated in Figure 6, bonding eliminates a difference in potential between objects. Grounding eliminates a difference in potential between an object and ground. Bonding and grounding are effective only when the bonded objects are conductive.

When two objects are bonded, the charges flow freely between the bodies, and there is no difference in their charge. Therefore, the likelihood of sparking between them is eliminated.

Although bonding eliminates a difference in potential between the objects that are bonded, it does not eliminate a difference in potential between these objects and the earth unless one of the objects possesses an adequate conductive path to earth. Therefore, bonding will not eliminate the static charge, but will equalize the potential between the objects bonded so that a spark will not occur between them.

An adequate ground will discharge a charged conductive body continuously and is recommended as a safety measure whenever any doubt exists concerning a situation.

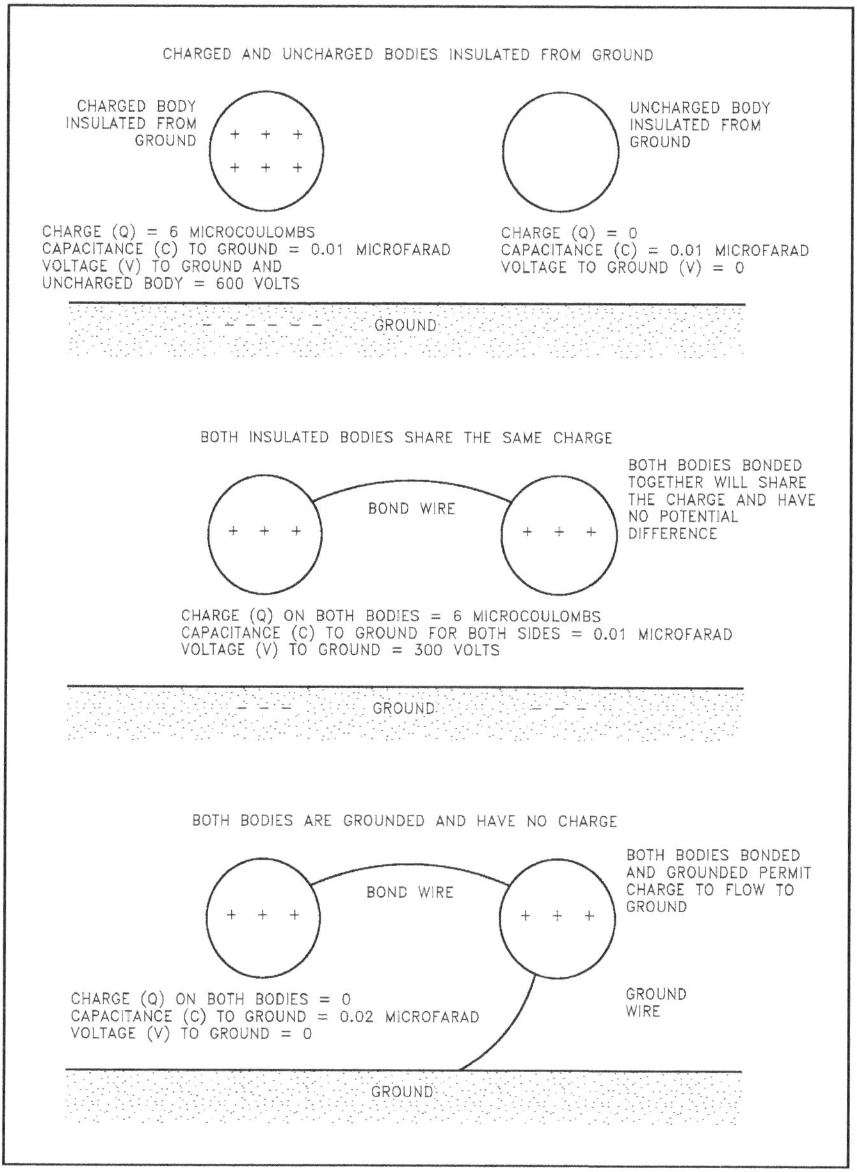

Figure 6 Bonding and Grounding

To avoid a spark from discharge of static electricity during flammable liquid filling operations, a wire bond should be provided between the storage container and the container being filled, unless a metallic path between the container is otherwise present.

Above-ground tanks used for storage of flammable liquids do not have to be grounded unless they are on concrete or on nonconductive supports. Ground wires should be uninsulated so they may be easily inspected for mechanical damage and should never be painted. Figure 7 illustrates grounding above-ground storage tanks.

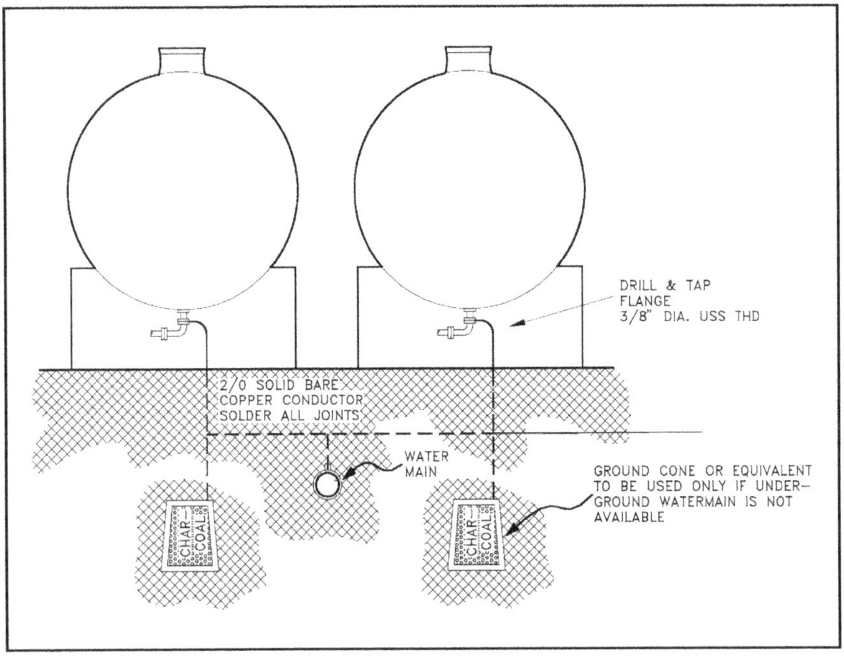

Figure 7 Grounding Above-Ground Storage Tanks

Petroleum liquids are capable of building up electrical charges when they (a) flow through piping, (b) are agitated in a tank or a container, or (c) are subjected to vigorous mechanical movement such as spraying or splashing. Proper bonding or grounding of the transfer system usually dissipates this static charge to ground as it is generated. However, rapid flow rates in transfer lines can cause very high electrical potentials on the surface of liquids regardless of vessel grounding. Also, some petroleum liquids are poor conductors of electricity, particularly the pure, refined products, and even though the transfer system is properly grounded, a static charge may build up on the surface of the liquid in the receiving container. The charge accumulates because static electricity cannot flow through the liquid to the grounded metal container as fast as it is being generated. If this accumulated charge builds up high enough, a static spark with sufficient energy to ignite a flammable air-vapor mixture can occur when the liquid level approaches a grounded probe or when a probe is lowered into a tank for sampling or gaging.

This high static charge is usually controlled by reducing the flow rates, avoiding violent splashing with side-flow fill lines, and using relaxation time, which allows time for the static charge to discharge.

When flammable liquids are transferred from one container to another, a means of bonding should be provided between the two conductive containers prior to pouring, as shown in Figure 8.

In areas where flammable liquids are stored or used, hose nozzles on steam lines used for cleaning should be bonded to the surface of the vessel or object being cleaned. Also, there should be no insulated conductive objects on which the steam could impinge and induce a static charge accumulation.

Nonconductive materials, such as fabric, rubber, or plastic sheeting, passing through or over rolls will also create charges of static electricity. Static from these materials, as well as static from the belts, can be discharged with grounded metal combs or tinsel collectors. Radioactive substances and static neutralizers using electrical discharges are also employed for this purpose.

Bonding and grounding systems should be checked regularly for electrical continuity. Preferably before each fill, the exposed part of the bonding and ground system should be inspected for parts that have deteriorated because of corrosion or that have otherwise been damaged. Many companies specify that bonds and grounds be constructed of bare-braided flexible wire because it facilitates inspection and prevents broken wires from being concealed.

Electricity becomes a source of ignition where flammable vapors exist if the proper type of electrical equipment for these atmospheres either has not been installed or has not been maintained.

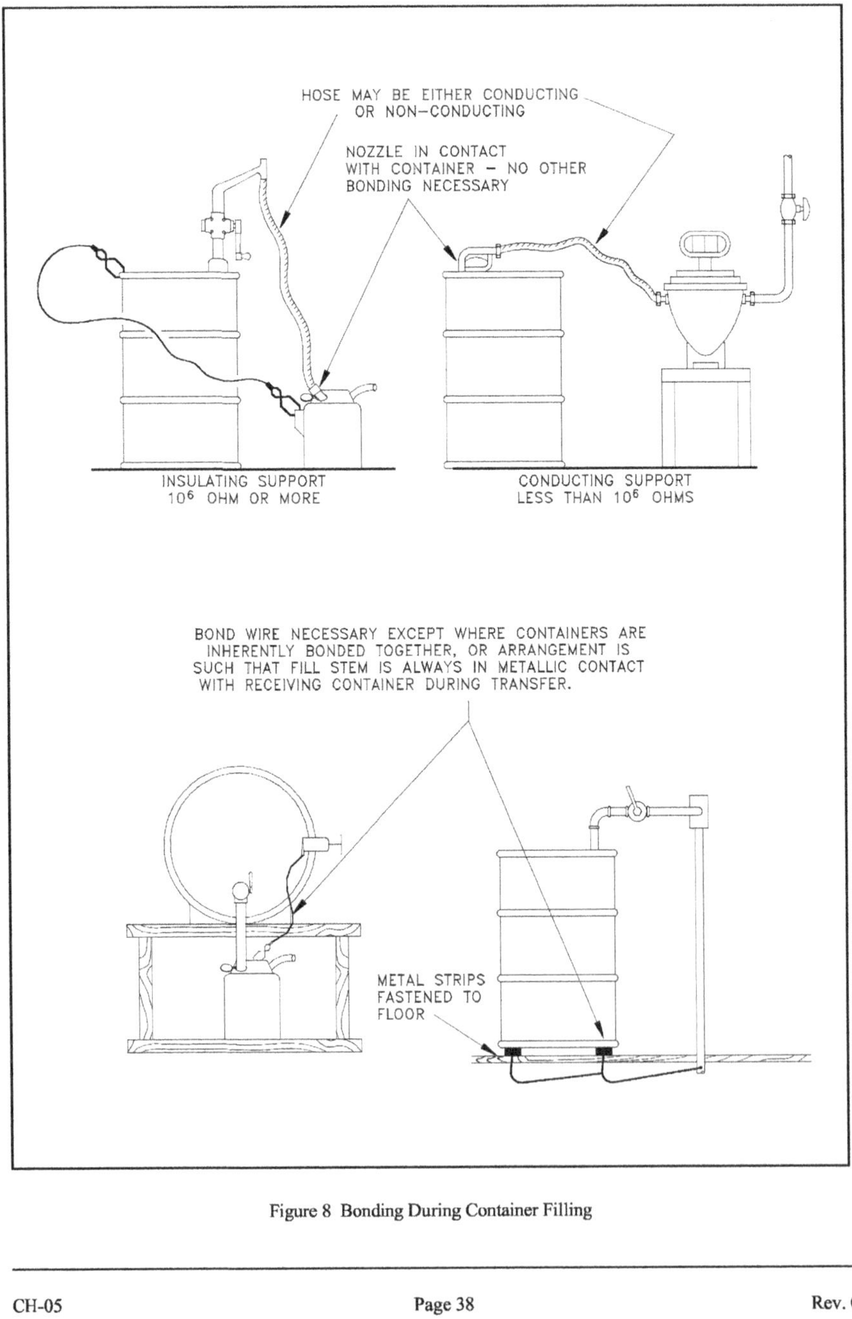

Figure 8 Bonding During Container Filling

A summary of reports of experimental evidence and practical experience in the petroleum industry shows that no significant increase in fire safety is gained by the use of spark-resistant hand tools in the presence of gasoline and similar hydrocarbon vapors. However, some materials such as carbon disulfide, acetylene, and ethyl ether have very low ignition energy requirements. For these and similar materials, the use of special tools designed to minimize the danger of sparks in hazardous locations can be recognized as a conservative safety measure. Leather-faced, plastic, and wood tools are free from the friction-spark hazard, although metallic particles may possibly become embedded in them.

Flammable and combustible liquids and their vapors may create health hazards from both skin contact and inhalation of toxic vapors. Irritation results from the solvent action of many flammable liquids on the natural skin oils and tissue. A toxic hazard of varying degree exists in practically all cases, depending on the concentration of the vapor.

Most vapors from flammable and combustible liquids are heavier than air and will flow into pits, tank openings, confined areas, and low places in which they contaminate the normal air, and cause a toxic as well as explosive atmosphere. Oxygen deficiency occurs in closed containers, such as a tank which has been closed for a long time, and in which rusting has consumed the oxygen. All containers should be aired and tested for toxic and flammable atmosphere as well as the oxygen level before entry.

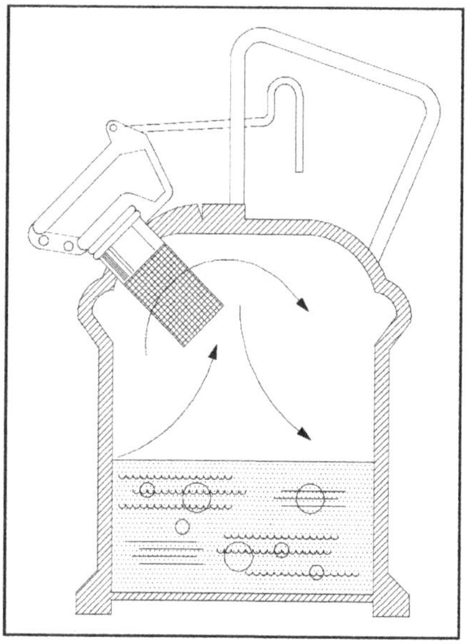

Figure 9 Storage Container With Spring-loaded Cover

Storage

Class I and Class II liquids should not be kept or stored in a building except in approved containers, as illustrated in Figure 9, within either a storage cabinet or a storage room that does not have an opening that communicates with the public portion of the building. The spring-loaded cover is designed to open in order to relieve internal vapor pressure. Quantities stored in such locations should be limited. They should not be stored so as to limit use of exits, stairways, or areas normally used for the safe egress of people. Neither should they be stored close to stoves or heated pipes, nor exposed to the rays of the sun or other sources of heat.

Losses by evaporation of liquid stored in safety cans at ordinary temperatures are negligible. Storage of flammable and combustible liquids in open containers should not be permitted. Approved containers for flammable liquids should be closed after each use and when empty. Warning labels should be removed from flammable liquid containers when empty (vapor free). Bulk Class I liquids should be stored in an underground (buried) tank or outside a building. No outlet from the tank should be inside a building unless it terminates in a special room, as illustrated in Figure 10.

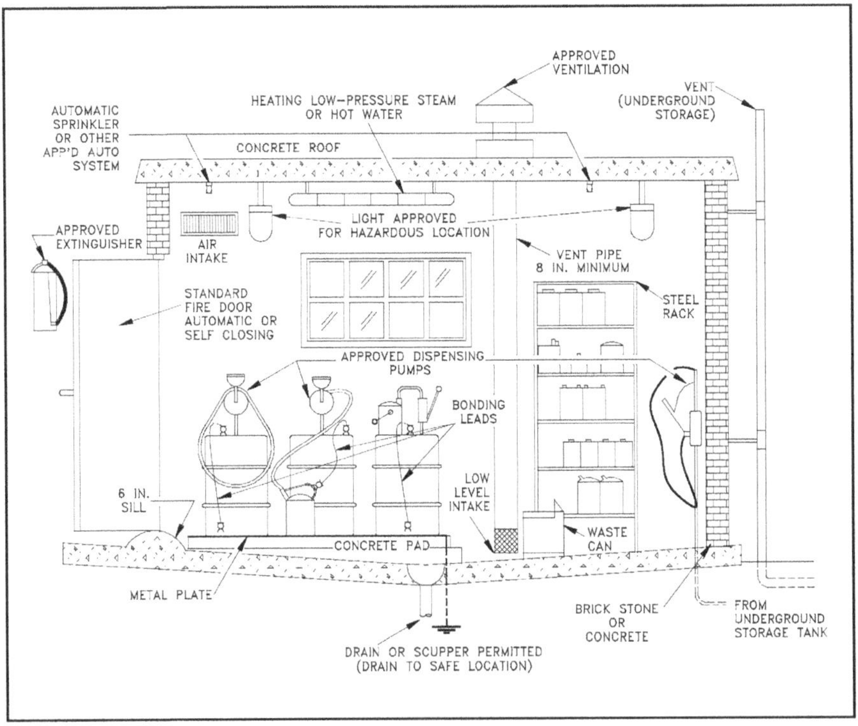

Figure 10 A Flammable Liquids Mixing and Storage Room

Vehicles used on plant property to transport flammable and combustible liquids in sealed containers should be designed to minimize damage to the containers.

When employees are filling tanks and other containers, they should be sure to allow sufficient vapor space (outage) above the liquid level in order to permit expansion of the liquid with changing temperatures. For example, gasoline expands at the rate of about one percent for each

14°F rise in temperature. Outage space for gasoline of two percent of the capacity of the tank or compartment is recommended, and permanent high-level markings should be installed.

Storage tanks should be provided with vents. Vent pipes of underground tanks storing Class I flammable liquids should terminate outside buildings, higher than the fill pipe opening, and not less than 12 feet above the adjacent ground level. They should discharge vertically upward, and be located so that flammable vapors cannot enter building openings or be trapped under eaves or other obstructions. Vent pipes from underground tanks storing Class II or Class III liquids should terminate outside buildings and higher than the fill pipe opening. Vent outlets should be above normal snow level.

Additional information concerning installation, protection, and spacing of storage tanks located above ground, underground, or in areas subject to flooding is beyond the scope of this course. If this information is desired, it may be found in the National Fire Protection Association Standards.

Flammable or combustible liquids in sealed containers represent a potential hazard rather than an active hazard -- the possibility of fire from without. By the same reasoning, inside storage rooms are undesirable. If they must be used, they should be isolated as much as possible, and located at or above ground level. They should not be located over basements and should preferably be along an exterior wall.

Every inside storage room shall be provided with either a gravity (low level intake) or a continuous mechanical exhaust ventilation system. Mechanical ventilation must be used if Class I liquids are contained or dispensed inside the room.

Storage cabinets have specific limits on the amount and class of flammable or combustible liquids that may be stored in them. They must be constructed and sealed so as to be fire resistant. Cabinets shall be labeled conspicuously -- FLAMMABLE-KEEP FIRE AWAY.

The most advisable storage facility is a separate building set some distance from normally occupied plant areas. The construction can be similar to that specified for inside storage rooms. The types and classes of flammable and combustible liquids stored will determine the best design to be used.

Summary

The important information from this chapter is summarized below.

Flammable and Combustible Liquids Summary

- The following terms are defined:

  Flammable liquid is one having a flash point below 100°F, and having a vapor pressure not exceeding 40 psia at 100°F.

  Combustible liquids are liquids with flash points at or above 100°F, but below 200°F.

- General safety precautions regarding the use, handling, and storage of flammable and combustible liquids include the following:

  The vapor-air mixture formed from the evaporation of the liquid poses a hazard; therefore, exposures of large liquid surface areas and sources of heat shall be avoided or prevented during handling or storage of these liquids.

  Accidental mixture of flammable and combustible liquids shall be avoided.

  Fill and discharge lines and openings, as well as control valves associated with flammable and combustible systems, shall be identified by labels, color coding, or both to prevent mixing different substances.

  All storage tanks shall be clearly labeled with the name of contents.

  Transfer lines from different types and classes of flammable products should be kept separate, and preferably, different pumps should be provided for individual products.

  When handling quantities of flammable liquids up to five gallons, a portable FM (Factory Mutual Engineering Corp.) or UL (Underwriters Laboratory) approved container should be used. The container should be clearly labeled.

  Smoking, the carrying of strike-anywhere matches, lighters, and other spark-producing devices should not be permitted in a building or area where flammable liquids are stored, handled, or used.

Flammable and Combustible Liquids Summary (Cont.)

- The reasons for and techniques used in bonding and grounding of flammable liquid containers are:

  Static electricity is generated by the contact and separation of dissimilar material. The principal hazards created by static electricity are those of fire and explosion, which are caused by spark discharges. A spark between two bodies occurs when there is a poor electrical conductive path between them. Bonding is done to eliminate a difference in potential between objects. Grounding is done to eliminate a difference in potential between an object and ground.

- Four sources of ignition concerning flammable liquids include the following:

  All sources of static electricity (induce sparks)

  Improper type of electrical equipment for these atmospheres or poorly maintained electrical equipment

  Flame-producing operations (smoking, grinding or welding)

  Improper hand tools that cause sparks

- The health hazards associated with flammable and/or combustible liquids are:

  Fire and explosions

  Skin irritation from the solvent action of many flammable liquids

  Vapors gather in low points or enclosed areas and cause a toxic environment

end of text.

CONCLUDING MATERIAL

Review activities:

DOE - ANL-W, BNL, EG&G Idaho,
EG&G Mound, EG&G Rocky Flats,
LLNL, LANL, MMES, ORAU, REECo,
WHC, WINCO, WEMCO, and WSRC.

Preparing activity:

DOE - NE-73
Project Number 6910-0021

www.KnowledgePublications.com

Made in the USA
Monee, IL
04 November 2026

67386418R00157